VISIONS OF QUALITY: HOW EVALUATORS DEFINE, UNDERSTAND AND REPRESENT PROGRAM QUALITY

ADVANCES IN PROGRAM EVALUATION

Series Editor: Robert E. Stake

Recent Volume

Volume 6: Telling Tales: Evaluation and Narrative edited by Tineke Abma

ADVANCES IN PROGRAM EVALUATION VOLUME 7

VISIONS OF QUALITY: HOW EVALUATORS DEFINE, UNDERSTAND AND REPRESENT PROGRAM QUALITY

EDITED BY

ALEXIS BENSON
University of Illinois, Champaign, USA

D. MICHELLE HINN
University of Illinois, Champaign, USA

CLAIRE LLOYD
University of Illinois, Champaign, USA

2001

JAI
An Imprint of Elsevier Science

Amsterdam – London – New York – Oxford – Paris – Shannon – Tokyo

ELSEVIER SCIENCE Ltd
The Boulevard, Langford Lane
Kidlington, Oxford OX5 1GB, UK

First edition 2001

Library of Congress Cataloging in Publication Data
A catalog record from the Library of Congress has been applied for.

British Library Cataloguing in Publication Data
A catalogue record from the British Library has been applied for.

ISBN: 0-7623-0771-4

∞The paper used in this publication meets the requirements of ANSI/NISO Z39.48-1992 (Permanence of Paper).
Printed in The Netherlands.

CONTENTS

LIST OF CONTRIBUTORS

Alexis P. Benson	University of Illinois, USA
Lynn Cortina	University of Houston, USA
Elliot W. Eisner	Stanford University, USA
David M. Fetterman	Stanford University, USA
Jennifer C. Greene	University of Illinois, USA
Delwyn L. Harnisch	University of Nebraska, USA
D. Michelle Hinn	University of Illinois, USA
W. Robert Houston	University of Houston, USA
Saville Kushner	University of the West of England, UK
Laura C. Leviton	Robert Wood Johnson Foundation, Princeton, USA
Claire Lloyd	University of Illinois, USA
Linda Mabry	Washington State University, USA
Linda Hamman Moore	University of Illinois, USA
Michael Quinn Patton	The Union Institute, Utilization-Focused Evaluation, USA

Sharon F. Rallis	University of Connecticut, USA
Gretchen B. Rossman	University of Massachusetts, USA
William R. Shadish	The University of Memphis, USA
David Snow	University of Illinois and Aurora Public Schools, USA
Theresa Souchet	University of Illinois, USA
Robert E. Stake	University of Illinois, USA
Judith Walker de Felix	Webster University, USA
Hersh C. Waxman	University of Houston, USA
Joseph S. Wholey	University of Southern California and U.S. Accounting Office, USA

PREFACE

Alexis P. Benson, D. Michelle Hinn and Claire Lloyd

ABSTRACT

Evaluation is comprised of diverse, oftentimes conflicting, theories and practices that reflect the philosophies, ideologies, and assumptions of the time and place in which they were constructed. Underlying this diversity is the search for program quality. It is the search for understanding and, consequently, is the fundamental, and most challenging, task confronting the evaluator. This volume is divided into four broad sections, each of which conveys a different vision of quality that include postmodern and normative perspectives, practice-driven concerns, and applied examples in the field of education.

Evaluation is comprised of diverse, oftentimes conflicting, theories and practices that reflect the philosophies, ideologies, and assumptions of the time and place in which they were constructed. Contemporary evaluators are, therefore, confronted with an ever-increasing number of approaches that tell them how to conduct evaluation. Each approach advances different goals to pursue (e.g. providing authentic information, measuring outcomes, making judgments of merit and worth, promoting utility, responding to stakeholders' needs, advancing normative agendas, etc.) and introduces different methods to guide practice (e.g. quantitative, interpretive, utilization focused, participatory).

Underlying this diversity is the search for program quality. It is the search for understanding and, consequently, is the fundamental, and most challenging, task confronting the evaluator. Quality is, after all, in the eye of the stakeholder. Hence, there may be as many definitions of quality as there are interpretations of experience. Moreover, a complex array of contingencies – in both local contexts and larger socio-political systems – shape stakeholders' particular experiences of program quality. Quality is also, inevitably, in the eye of the evaluator. Evaluators' definitions of quality – the understanding they seek – is informed by their ideological orientations and epistemological preferences. For some, quality is synonymous with merit, worth, or other evaluable properties; for others a sense of quality emerges as they seek a holistic understanding of program conditions; while still others do not so much seek to understand quality, but to advance a particular dimension of quality (e.g. progressive social change).

The approaches employed to guide practice reflect the evaluator's definition of quality. The goals evaluators pursue, the issues they focus upon, and the questions they ask determine the understanding of quality they seek. Evaluators may investigate the relationship between a program's goals, objectives, and outcomes; they may construct contextualized understandings of the experiences of diverse stakeholders; or they may create a normative knowledge base that can challenge traditional knowledge claims and transform the assumptions that guide practice. The methods and procedures evaluators employ to pursue their goals determine the nature and dimensions of the quality they find. Broadly speaking, methods choices reflect the extent to which evaluators rely on objective measures of prescribed criteria to analyze program conditions or on their interpretive ability to make sense of the interrelated contingencies that shape stakeholders' program experiences.

The way evaluators represent their understanding determines the quality they reveal. Among the many representational forms that may be employed to communicate quality, evaluators can synthesize findings on prescribed criteria to generate authoritative judgments of program merit or worth, describe the experiences of diverse stakeholders' to provide new and deeper understandings of program conditions, or provide meaningful renderings of stakeholders' voices using a range of techniques (e.g. stories, dramatic readings, dialogue, public forums, etc.).

The fundamental intention underlying all efforts to understand and represent quality is the desire to do no harm and, following from this, to do some good. Evaluators want to foster program improvement, to advance stakeholders' understanding of their program, to enhance the program's capacity to serve the public interest, and so on. In light of these intentions, evaluators want their understandings to be more authentic than artificial and their representations to be more substantive than superficial. Thus, they are compelled to search for better ways to uncover and understand their visions of program quality and better ways to represent these understandings to stakeholders. Moreover, as theorists in evaluation, and other disciplines in social inquiry, continue to gain philosophical and ideological insights that inform practice, evaluators are confronted with new dimensions of quality to explore, new procedures to employ, and new representational challenges to confront.

In this volume we wanted to explore different visions of program quality. We asked a range of evaluators to define quality and consider what this means for the way that they think about and conduct evaluation. The collective efforts of all who responded provide a broad theoretical perspective of quality. By exploring the ideological concerns and epistemological assumptions that support their visions, the contributors provide insight into the many different ways quality can be defined. By placing these perspectives within an evaluative framework, they connect their definitions of quality to the approaches that define the landscape of the field.

Many of the contributors also describe their efforts to understand and represent quality in everyday program contexts. These descriptions demonstrate how different evaluative approaches create particular understandings of quality and how different representational forms and styles communicate particular facets of quality. Moreover, in juxtaposing their ideological and epistemological preferences with the realities of evaluative practice the contributors indicate the vision of quality that is most effectively pursued given both the nature of the program and the socio-political context it inhabits.

We have organized the volume into four broad sections, each of which conveys a different vision of quality. In so doing, we do not intend to oversimplify conceptualizations of quality, or overlook the diversity between definitions, but to make sense of the range of the assumptions and purposes that guide contemporary evaluative practice. We hope this will help readers locate themselves epistemologically in an increasingly diverse field and make informed decisions about the vision of quality they will pursue given the theoretical and practical possibilities of different social program contexts. This will, we believe, contribute to the development of new visions of program quality that can facilitate better understandings and more meaningful representations.

The chapters in the first section are imbued with a distinctly postmodern sensibility. The contributors offer a vision of quality that recognizes the complex social world programs exist within and contribute to. They explore ways to construct holistic understandings of program quality by bringing stakeholders' diverse, possibly conflicting, local experiences into contact with larger discursive, social, and material systems. The contributors also consider the representational demands engendered by the epistemological characteristics of postmodern understanding. In particular, how to establish the authenticity of evaluative knowledge claims that are characterized by a plurality of complex inter-relationships which remain partly unresolved, and how to communicate authenticity to stakeholders in the absence of absolute, authoritative meaning.

The second section reflects the normative turn that has gained momentum in evaluation over the past decade. The contributors provide insight into the following issues: the social action agenda that shapes their visions of program quality; the distinctive methodology they employ to create normative understandings of program conditions; and the alternative representational forms they use to initiate a process of progressive social change. Many contributors also provide practical examples of evaluation grounded in an emancipatory agenda. These accounts reveal the relationship that develops between the evaluator and program stakeholders as they assume joint responsibility for the construction of a normative knowledge base; the barriers evaluators may encounter as they ensure diverse stakeholders are fully and fairly engaged in "rational social discourse"; and the benefits that accrue as a result of a dialogic process of knowledge construction.

The third section focuses on the practical issues and concerns that drive social program evaluation. Although the contributors present different approaches to guide practice (e.g. utilization evaluation, performance based management, descriptive valuing), their work is grounded in an effort to address the information needs of program personnel in order to provide useful information that will facilitate program improvement. In prioritizing these objectives, the contributors demonstrate how design decisions and methods choices are often determined more by the practical realities of the program context (i.e. the nature and size of the program, the larger forces the program comes into contact with, the needs and values of diverse stakeholders) than by the evaluators' particular ideological and theoretical orientations. They provide insight into several fundamental issues evaluators must consider to ensure the provision of useful information that will enhance program functioning. Particular attention is given to the different roles evaluators can assume to contend with multiple views of program quality in order to facilitate the application of the evaluative information.

The final section continues many of the themes relating to evaluation theory and practice introduced in preceding sections, but it does so through a consideration of evaluation in educational settings. The contributors challenge the standards-based accountability models that continue to dominate educational evaluation. They suggest that the universal, normative knowledge claims associated with these models are inconsistent with a vision of quality that values student diversity and recognizes the complex competencies associated with educational leadership. In light of this inconsistency, the contributors discuss how evaluators can avoid perfunctory understandings of program quality by moving beyond the circumscribed dimensions of traditional evaluative models and allowing understanding to emerge through interactions with diverse stakeholders. They also reject traditional representational formats and present a range of alternative forms and styles that can be employed to transform entrenched and intractable "ways of knowing."

Alexis P. Benson, D. Michelle Hinn and Claire Lloyd, December 2000.

SECTION I

1. REPRESENTING QUALITY IN EVALUATION

Robert E. Stake

INTRODUCTION

Among the responsibilities of professional evaluators is representation of the quality of the evaluand. Often, after an appropriate description of program performance, program value is only alluded. Often the evaluand is misrepresented because only a single facet or perspective of value is featured. Representing quality is very difficult, and the immodest evaluator is reluctant to admit that high confidence in the representation is not warranted. Of course there are multiple definitions of quality. I will speak of different definitions and representations here, and my fellow authors will provide others, perhaps more compelling definitions, in the chapters to follow.

QUALITY AND REPRESENTATION

We have two constructs here, both central to evaluation as a form of human discourse. Neither construct is easily defined. First, *quality*, perhaps the most essential idea in evaluation. To evaluate is to recognize value, to recognize quality. Evaluating is *not first*, the measuring of certain criteria. And *not first*, facilitating decision making. Evaluation is *first*, the search for goodness and badness, for merit and shortcoming, for quality.

The second elusive construct is *representation*. In evaluation, we not only need to *discern* quality but to *convey* the sense of quality to others. And then

Visions of Quality, Volume 7, pages 3–11.

ISBN: 0-7623-0771-4

also, we need to comprehensively identify the evaluand, the thing with which quality is experienced. Thus, to our audiences, we represent twice: we represent quality and we represent the evaluand. It makes little sense to provide a sharp indication of quality if the object evaluated remains uncertain.

Quality as Construct

We can think of quality as a property of the evaluand or as a construction of people who experience the evaluand. To be of service to a diverse set of stakeholders, we evaluators need to do both, to present quality as inherent in the program evaluated and as experienced by people. To be of most service to society, by my political philosophy, we should honor human perception by favoring the concept of value as a construction over quality as a property. I will go so far as to say that adherence only to the concept of quality as some intrinsic property of the evaluand is a misrepresentation of quality.

I have become persuaded only lately that, by our definition of quality, we support or oppose different modes of evaluation. As we go to the field and as we write our reports, the more we treat value as a property, the less we honor program stakeholders. The more we treat value as a situational construct, the more open we are to honoring program stakeholders. Or in other words, the more we support a single, traditional, external, rational definition of quality, the less we honor the definitions of quality by those most affected by the policy and program.

If quality is not so much to be discovered in the program itself but to be understood in terms of experiences of participants and constituents, then we need models or theories of *stakeholding*. Whose program is it? Who counts? I don't know how to do this formally, but we all do it intuitively. Should we do it more rationally? Sometimes surely we should, but it is more important to have critical, meta-evaluative mechanisms along the way that keep raising the questions. Who counts? How are we representing quality?

Egalitarian Definition

Everybody constructs the quality. I think we should take the egalitarian view that whosoever is moved, then there is quality. It doesn't mean that we have to admire everyone's quality, but we should hold it up for examination.

We grandparents observe the children performing. For those things we did not expect ours to be able to do, we are moved by the quality of it. The experts, the teachers often see in the performance less quality than they are looking for. But it is quality, for us who are moved. I think evaluators need to find the audiences that are moved, thrilled, perhaps repelled, as well as to have the data scrutinized by blue ribbon reviewers. It is quality for those who are moved, whether or not it is for the rest. Averages are not important. Standardization serves us poorly. We find life enriched by quality, however hidden it may be to everyone else. When we face standards that demean our enchantment, quality of life is lowered, not just for one, but for all.

Derivative Standards

I would like to tell you what I perceive to be the origins of quality. Collectively and personally, it comes out of comfort, out of contentment, out of happiness. Quality emerges not as a property of some object but out of experience with that object. No matter how complex the characterization of quality, its origins are rooted in experience.

As we express ourselves, we invent representations of quality. The resonance of quality becomes gradually less personal, more formalized. We invent criteria and standards, and they evolve. Back and forth, experience and verbalization interact. We speak of Grade A milk, not only because we like its taste but also because we cherish the health we associate with eating well. The cherishing is basic; the verbal description is a derivative definition of quality. Grade A is a derivative status of milk.

A teacher becomes "Teacher of the Year", a model, an object of comparison. In summer school, a Chicago third-grader "fails the Iowas", again, and must repeat third grade, again. For wristwatches, Rolex is a standard of quality, a derived standard, useful for comparisons. For professional critics in the arts, instances such as Sousa, *Old Man and the Sea*, Degas, and *Death of a Salesman*, serve as standards. A useful standard may be only indirectly tied to immediate experience.

Sometimes we act as if these derivative standards of quality are basic, that they get at some essence. We recognize that compared to the world's experience, most personal experience is primitive and biased. But the world is not always the most relevant. For program evaluation, definitions of quality need regularly to be traced back to how people are pleased, back to origins in experience.

Social Construction

As we rationalize our pleasures, as we come to understand our contexts, we see quality in an increasingly disciplined sense. Each of us develops a personal discipline. The discipline we develop for ourselves is not identical to the disciplines others develop. Quality becomes defined both ways, once through our own experiential participation, and again by beholding the standards of others.

For us constructivists, quality doesn't exist until people declare it so. And people declare it when struck by the exquisite, when moved by the encounter.

Probably simultaneously, we are aroused, we are emotionally involved, and we see it as something special, something aesthetic. The quality of teaching was not there until we saw it. Or until someone saw it and pointed it out to us.

At the same time we are developing our constructs of quality, others are developing theirs too. Much of what we construct, we construct together. Aesthetic quality is a social construction. The roots are in experience but the characterization, the representation of quality, often becomes far removed from experience.

Of course, some people do not experience what others experience, and thus a side of quality is hidden from them. Yet the meaning of quality is formed by everybody, including those who do not experience it. If most of the stakeholders *do not* experience the merit of the program, the program has less quality than if most do. The quality of a social venture needs to be seen both as a function of the total experience of its stakeholders and as the experience of individuals and subgroups.

Preserving Independence

There are critiques, negotiations, political maneuvers, perhaps revelations. Agreements are struck and certain styles, sounds, movements and arrangements become standards. Ceaselessly, authority exists, ranking programs and personnel. The same is true for products from cheeses to chiffoniers. Personal experience is squeezed by the experience of others. The loudest and most eloquent advocate their criteria, and by their persuasions, maintain their roles as critic and judge.

Quality does not and should not reside alone in expert declamation. Standards need to be tempered, and sometimes replaced, by personal sensitivity. The reality of quality does not depend on the ability to explicate it. Whether participant or spectator, whether producer or consumer, whether client or evaluator, the individual has to make a choice.

The evaluator has to decide how much to rely on inside and outside voices. Sometimes the press of society makes it very difficult to exercise independence. Evaluators, like most people, cannot be satisfied with conventions of disciplined judgment that conflict with their own personal discipline.

Personal Disciplines

As we come to understand our pleasures and the values around us, we see quality in an increasing aesthetic sense. When we encounter high quality in art and music, the exquisite, many are moved; many are thrilled; many are in awe. Many are not. The construction of merit is rooted in such experiences.

There is quality also in the mundane: in chairs, in pencils, even in paper clips. Are we thrilled by a paper clip? No, but we are dismayed if the edge cuts our finger, if the clip doesn't close again after use, if the price of clips has doubled. Low quality is a mirror of our dismay. How can the price of paper clips or any other property be an aspect of quality? It is not, unless the user sees the evaluand of different quality because of it. All this does not mean that customer satisfaction and quality are the same thing. The customer can perceive the quality high and the purchase poor.

If the successful program is found to burn out its staff, the stakeholders' perception of program quality may change. Sometimes a free program is valued less than a high-cost program. Costs may temper the sense of goodness. The evaluator does not need to know the formula. Or to validate it. According to their personal disciplines, stakeholders form ideas of merit including, consciously and unconsciously, whatever ingredients life and reason teach them. Quality is a construction of the stakeholder and, some of the time, needs to be treated that way.

But would it not be misrepresentation to rely only on the feelings and comprehensions of the stakeholders? It is the responsibility of evaluators, I think, to draw implication further than stakeholders are aware, to express value partly as what we the evaluators judge, is in the stakeholders' best interests. We are not professional evaluators unless we represent quality as others see it and as we see it.

Disciplines of Quality

The recognized disciplines of the arts and of all objects and performances of quality are not democratic institutions. It is not one person, one vote. As a result of some ordination, a certain ownership, various persuasions, a relatively few people exercise control over most formal disciplines.

Most of us in evaluation honor an academic tradition that says that quality should be rooted in scholarship. It is a tradition saying that those who study these matters most should have the most to say about merit, about quality, about goodness. The tradition holds that we should use the conceptualizations of those who care most about meanings, those who refine the meanings (sometimes we call them *connoisseurs*) those who articulate best, should have the final say. That tradition is strong among us. And, as elsewhere, we should be skeptical about it.

Let us make a distinction between identifying the grounds on which quality might rest and declaring the standards by which quality should be determined. Thus I say, description, yes; prescription, no. We should listen carefully to the connoisseurs but we should judge program quality first in terms of the experience and well-being of the stakeholders. What is in their best interest? I would have us represent value by seeing it less a property and more a stakeholder construction.

Correlates of Quality

Many evaluators use a certain set of outcomes as their measure of quality. Some look pointedly for goal attainment, others for efficiency, still others for utility. In a context we know well, it becomes second nature for all of us to see quality upon encountering the searched-for characteristic. The Japanese have a saying: "It may be rotten but it is red snapper". We come to recognize the imprimatur of worth, the signs of success. We don't have to be joyous or aesthetically moved any more. Evaluators are expert at recognizing the surrogates for quality.

Evaluators sometimes indicate quality by referring to a formal standard, a special comparison. In our work it is usually a verbal standard. The standard in some way traces back to original encounters in which the standard setters were moved. All standards of quality thus are derivative. One can use a check-list of quality to evaluate a project, but the checklist needs to be validated against experience.

With increasing familiarity with the evaluand, the perception of quality changes. We adapt to the presence of quality, knowing it increasingly relies on cognitive rather than conative grounds. Some new objects are so closely associated with objects already known that we attribute the quality of the one to the other. I do not say that a school is without quality if it fails to thrill us or repel us. I do say that school quality is often best thought of as a product of personal experience.

It will sometimes be useful to draw upon the experience of experts, of connoisseurs. It is important to know their perceptions of quality, their formulas, their standards. Certainly, theirs are not the only important perceptions. Teachers, students, and other stakeholders, all have important perceptions. The usefulness of connoisseurs often is their ability to provide a language for comprehending the quality that others recognize but cannot communicate. It will sometimes be useful to draw upon the skills of quantitative analysts, but measurement is not the ultimate representation. There is no obligation for the evaluator to aspire to some weighted synthesis of various images. The analyst is an expert at identifying factors insufficiently discerned. With expert help, a panorama of awarenesses of quality is there for the evaluator to discern.

Representation

Now to speak more technically of representation. An evaluation report is a house of representatives. It is filled with representations of the evaluand and of value. For example, the evaluation report at the close of every World Bank overseas development project must explicitly say whether or not the work of the Bank was satisfactory or unsatisfactory. It at least briefly describes the works provided, e.g. the irrigation, the training, the entrepreneur-support, the schools. It has a discussion of objectives, actions taken, impediments to full implementation – all representations. It is important for every evaluator to let others know the nature of the evaluand as *perceived*.

To represent is to stand for, to be a surrogate for. Words represent ideas. A name is a representation. But a name tells so little. It is the nature of a representation to be incomplete. Some representations will be misrepresentations only because they are so terribly incomplete. We sometimes seek symbolic representation that can be quickly comprehended, names, acronyms, logos, indicators. And at other times, we seek representations that portray something of the complexity and the many contexts of the evaluand.

We have colleagues such as Patrick Grasso, Mike Hendricks, Andy Porter, and Michael Scriven who, at least at times, urge the reduction of representation to the fewest indices. And we have colleagues, such as Tineke Abma, Elliot Eisner, Jennifer Greene, and Linda Mabry, who urge experiential accounts, narratives, stories to convey the sense of the evaluand as a living phenomenon. And those of both schools oversimplify. They cannot but leave out important qualities. They displace the complex with the simple. Each representation cannot but be misrepresentation.

Still, incompleteness is less a worry than obfuscation. Some representations are just plain confusing. Our colleagues in quantitative research have relied much on the random sample for representing populations. Randomness allows precise indication of error variance and confidence interval but does not provide representativeness. Take it a step further. No sample size one is representative of any but itself. No case adequately represents others. No legislator represents the people of her district. No member of the advisory committee represents people having the same demographic description: male, veteran, teacher, connoisseur. Representation is a slippery concept.

Evaluation as Interpretation

Representation requires interpretation. Measurement is easy. Interpreting measurement is difficult. Evaluators are challenged to give meaning to events, relationships, needs, and aspirations, far more complex than their data can convey. The representations created by the evaluator seldom adequately mirror the things observed. All languages fall short, but some more than others. Analytic specification and operationalization are sometimes useful, not because they get closer to truth but because they can be grasped. The evaluator composes representations of the evaluand and of its quality. Drawing from all the languages of art and science, the evaluator describes program activity, staffing, contexts, coping, and accomplishment, sometimes hoping to provide a vicarious experience, sometimes hoping to create descriptors well beyond personal experience. Within the descriptions are the clues to quality. A good description cannot but tell of quality. But the evaluator goes further, to summarize the conclusions of merit, grounds for the conclusions, and implications of the findings of merit. Evaluators struggle to relate quality to the issues, to the complexities, to the implication for action. Formulas for representing quality are no less complicated than formulas for writing poems, novels, and symphonies.

The reason for failure of experimental designs for evaluating our reforms is not so much the lack of a suitable control. Failure is assured because the "treatment" cannot be adequately represented. In education and social service, the treatment is never a single change element but a vast array of changes. There are so many treatments in the one experiment that good representation is impossible, leaving attribution confounded, and the thoughtful evaluator bewildered as to what might have caused any increase in quality of outcome.

The evaluator constantly seeks compromise between accuracy and comprehensibility. Both the case study and the performance indicator mislead. Both suggest meanings and precision not found in the object itself. Metaphors serve partly because they advertise their imperfect fit. Fallacies in representation need to be advertised. Lacking a concept of standard error of measurement, we qualitative evaluators need to seek our own ways of informing the reader of the confidence that can be placed in our representations.

Deliberate Misrepresentation

In these times, the most sophisticated representations are not those of art and the sciences but those of advertising. Lying is not a new invention, but the quality of lying has been greatly raised. Sound-bite campaigning. Image making. Audience-capturing commercials.

In evaluation, what is the quality of our representations of educational value? Are our executive summaries the equivalent of sound bites? Do we bury the bluster and insensitivity of the program in small print and impenetrable performance indicators? How frequently is the evaluator "party to the lie"? Can the evaluator avoid looking for those representations that will bring more support for disadvantaged children? As we analyze and interpret data in the reality of the schools, do we lean toward accuracy or advocacy?

It is regrettable that the invalidity of representations of program quality is so little challenged. There are those who object but they have little voice, little counter representation. The image-makers want to show program performance more than honest reporting.

Is it different with our other clients? Which is more important to them, to know the quality of their program or somehow to obtain representation or misrepresentation that creates the impression that the program is effective? Our clients feel differently at different times. Sometimes, simplistic indication of gain looks better than complete analysis of effort. Misrepresentations can be of great value.

Representation of Quality

Evaluators will continue to be caught up in the advocacies of the programs, in the imagery of the marketplace, and in the compulsion of the accountability movement. In order to demonstrate good work, every modern department and agency is caught up in demonstration of performance. They are required to represent quality of impact by identifying aims, strategies and outcomes. Here and elsewhere, a media-saturated demand for representation far outruns our ability to represent.

The needed qualitative "representation of misrepresentation", that counterpart to the standard error of measurement, is critical meta-evaluation, the periodic and climatic critical review of process and product. We can be more confident in the messages of evaluation as we raise challenges to representation. Such procedures are not yet systematized, but they already are part of the better evaluation studies.

Progress may depend on further evolution of the meaning of quality in program evaluation. Does it help to separate quality as a property from quality as construct? The chapters that follow summarize the purview at century's beginning and expectations for interpretation still to come.

2. WHY FINDING NEW FORMS FOR REPRESENTING QUALITY MATTERS

Elliot W. Eisner

My comments in this essay are largely theoretical and speculative; they are an exploratory effort to identify the ways we might think about how we can represent what we care about in education. The issues I will examine are not resolved. Two issues that need attention are the meanings we assign to the term quality and because we are interested in the way in which meaning is conveyed about quality, the systems we use to represent it.

Perhaps the place to begin is to acknowledge that the term "quality" is systematically ambiguous. On the one hand quality refers to what we value, as in the phrase, "That was a quality performance". On the other hand we use the term quality to refer to the sensory features of a process or product, as in the phrase, "This piece of wood has a smooth quality". This latter use of the term quality is like the term "quale" as used in philosophical discourse. Quale refers to feel, quanta refers to amount. In psychology the German gestalt psychologists talked about "gestalt qualitat" or form quality; that is, the perceptual features of figure-ground relations. To complicate matters even further the term quality can refer to both the feel of something and the value we assign to it, as in the phrase, "He seems edgy". Edgy, a metaphor, describes both a quale and expresses a value judgment.

The point of my remarks is to explore the ways in which we in education now represent quality and how it might be enriched. In my remarks I will be

Visions of Quality, Volume 7, pages 13–18.
2001 by Elsevier Science Ltd.
ISBN: 0-7623-0771-4

addressing two uses of the term, quality as a value judgment and quality as a representation of the feel or look of something.

Before proceeding with the exploration, a word needs to be said about why such considerations are important. Why be concerned with how quality in education is represented? Why care in light of all the important problems in the field?

There are several reasons. First, attention to how we represent quality is a neglected issue in education. We use what we use – grades and scores – and that means we use what we have been using to represent quality. Now neglect is not a sufficient reason for paying attention to something. If what is neglected is trivial, it doesn't matter if it's neglected. But the way we represent quality is what the public receives that tells them, for good or for ill, about the quality of schooling being provided and it tells them how well their children are doing. What could be more important?

Second, we need to know about the quality of the curriculum we provide, both in vitro and in vivo. We need to know about the quality of the teaching our children receive, we need to know about the quality of the environment in their school, we need to know about the quality of what they are learning. How those qualities are represented influence what we can know. What if the forms of representation we use limit in deleterious ways what we look for in schools? What if we can invent ways to represent quality that would enrich rather than diminish what we could say about the outcomes of our efforts? Might this not enlighten educational policy and perhaps contribute to better practice? These seem to me reasons enough to see where the exploration of the representation of quality leads.

Let's start with a distinction between digital and analogue systems for representing quality.[1]

Digital systems are systems designed to reduce ambiguity by employing rules for the organization of elements within its system, rules intended to constrain options for interpretation. Consider the use of number and the process of calculation. Given a system for treating numerals, those using that system, if they share the rules of the system, will arrive at the same numerical conclusions. Digital systems also have rules of equivalence that make the substitution of the elements possible without loss or alteration of meaning. Thus, 4 + 4 = 8 could be written: eight, VIII, IIIII III, 7 + 1, 6 + 2, 10 – 2, or even "ate"! The rules of equivalence make possible constancy of meaning because the rules of the system prescribe the ways in which the elements are to participate.

Examples of digital systems include the use of number for ranking, measuring, they include the use of visual devices such as pie charts, histograms, and, depending on the level of linguistic precision possible, literal descriptions of states of affairs.

Our grading system in schools, our ranking of students, and our use of transcripts for recording the quality of a student's performance employ digital systems for representing quality. The use of letter grades make it possible to calculate grade point averages because we have a system for transforming letter grades into numbers and then into numerical averages. Digital systems state meaning, but as we shall see analogue systems express meaning.[2]

Analogue systems have no explicit rules of equivalence. Analogue systems employ both iconographic and the expressive capacities of form that make it possible to infer from those forms something about the phenomena one wishes to know. Consider narrative, especially literary narrative. Literary narrative provides the reader with a form through which to secure a sense of the place or the people involved in a situation. Through literary narrative we experience the quality of the place, that is, its feel, but we can also learn what value the writer assigns to it. Put more simply, through the story we learn what the place is like and how it is appraised. Simply recall the film *Schindler's List*. What we encounter here is both quale as feel and quality as value. Without the story, the images, and the music we would know less of what life in concentration camps was like. In this example the use of both sense and reference advance our understanding. Sense gives us a feel for the situation and reference, represented through conventional signs, denotes. Together they express an array of values.

My line of argument here is that *Schindler's List* and other analogue forms are, among other things, representations of quality in both of the senses I described earlier. Because these forms describe qualities, we can ask if what they do through film has a potential role to play in disclosing quality in education, in schools, in classrooms, and in teaching and learning. And if they do, what do they provide that is not now provided?

Analogue systems are by no means limited to narrative. Analogue systems include pictures, exhibitions, performances, restaurant menus, and descriptions of the qualities of wine – in both senses of the meaning of quality. Analogue systems require inferences by those who "read them". They work by innuendo, they exploit the expressive capacities of form, they traffic in metaphor, they suggest and imply rather than claim or state. They are used because innuendo, connotation, and metaphor often convey meanings that literal statements cannot reach (Eisner, 1998).

The educational question is this: Is the use of analogue systems as a way to represent quality in, say, teaching and learning feasible? And is it likely to be useful? What we now use with few exceptions are grades and scores. Grades are the summaries of a mix of types of data – test scores, observations of student performance, assessment of the quality of written material and the like. All of this information in its rich particularity is reduced to a letter, sometimes modified by a plus or minus. Grades are then further reduced to a GPA and in the process even more information is irretrievably lost. There is no way in which anyone given a grade can reconstruct the basis for its assignment. In fact, two students taking the same 100-item test can get the same total score for exactly the opposite performance on the same test. The first student gets the first 50 items correct and the second 50 wrong. The second students the first 50 wrong and the second 50 right. Their total test score is identical, but their performance is not. The problem we are coping with is one of finding forms for representing quality that have an important and distinctive contribution to make in edifying a reader about the quality of a performance or situation.

There are some examples of the use of analogue forms of representation for disclosing quality that have to do with representing and appraising schools. One of the most vivid examples, perhaps, is found in Jonathan Kozol's (1991) *Savage Inequalities*. Here we get in living color portraits of what schools serving children of color provide to those children. Such a rendering informs us in ways statistics never could about the living hell those children had to endure and by implication our own national complicity in their lives. We have similar functions performed in Alex Kotlowitz's (1991), *There Are No Children Here* and in Frederick Wiseman's (1980) classic film, *High School*. What we have in these works are representations of quality in both of the senses I mentioned earlier; we experience its sense and appraise its value. Such renderings through an analogue system are also made possible through video, poetic displays, and exhibitions.

But is such an approach to the disclosure of quality appropriate and feasible for individual children? What do we do when the so-called "unit of analysis" is the child?

Let us at this early stage of the inquiry postpone attention to matters of feasibility. Let us ask rather whether it can be done. It seems to me that the answer on the basis of experience is "yes". We do it now, particularly when the stakes are high. Consider the way we represent quality when we review new potential hires at the university. If your university is anything like mine we first review letters submitted in response to ads in the newspaper. This is the first representation of quality we encounter, one provided by the candidate. Second, we cull from that list of applicants a handful whose personal representations we find impressive. We then as a committee deliberate. We read materials submitted or articles published. We invite a short list to visit for interviews and a job talk. Here too quality is being represented and in countless ways: personal drive, clarity, imagination, a nose for the significant. We read letters of recommendation, another representation of quality. We deliberate and try to make a case that is compelling for each. In this effort we represent quality by writing a "story" that conveys to the faculty at large our recommendation.

In none of this critical work do we ask our candidates to take a quantitative test or submit a transcript or any other digital form of description. In fact, in the selection of faculty, one of the two most important decisions we make, we put stock in the deliberative, not the calculative process. It is not the number of articles someone has published, but their quality. Furthermore, if we felt comfortable using something like the faculty equivalent of the GRE's to make our selection, the process would be far less arduous – but at Stanford, at least, it wouldn't fly.

So what does all of this discussion about representing quality mean for our schools – and in the United States we have 108,000 of them? It seems to me it means this. It means that we can explore the ways we represent quality by inventing alternatives. The use of portfolios and student exhibitions, with all their difficulties, are steps in the right direction. So is performance assessment when handled well. When supported by text, portfolios can display much that cannot be disclosed by number or letter.

The University of California at Santa Cruz, for example, has for years used narrative descriptions of student performance in lieu of grades. It's true that that these descriptions, often several paragraphs in length for each course for each student, are more time consuming to write and read but they also provide a more replete picture, that is, a fuller picture of not only what the student has accomplished, but how the student worked. Equally as important, such descriptive procedures remind the faculty to pay attention to the student in ways that under ordinary circumstances an instructor can afford to neglect; the expectation that a narrative form of representation is to be used to describe and appraise can effect the way instructors pay attention to students.

There is a down side. Analogue systems are likely to be regarded as subjective with all of the liabilities, legal and otherwise, that subjectivity entails. There is another down side. In addition to the time required to write and read narratives, narratives make comparisons among students more difficult. When every one has a GPA and a GRE it is easy to make comparisons among the "units of analysis". But there is a downside to digital systems as well. Scores, for example, breed an unwarranted sense of confidence in levels of precision that may not be there. When you are dealing with an analogue system the distinctiveness of each student becomes a salient consideration and when this happens incommensurability among students increases. Is this a problem? Should our traditions for representing quality define our aims? How important is comparability?

It is easy I think to underestimate the effects of the ways we now represent quality. Our current procedures of grade giving and score assignment, of check-off forms and rank ordering have profound consequences for defining the values that animate our schools. We love league tables! The systems we now use are consequential and some of us do not value their consequences; they neglect what is distinctive, they are excessively data reductive, and they make it impossible to know what went into the determination of the grade, they privilege the affluent, they distract attention from what is intrinsically valuable. Yet, up to now we have not been able to design robust alternatives. Perhaps we never shall. Perhaps the social system within which our schools function needs losers in order to have winners. Perhaps we use ranks to sustain social privilege. Whatever the reasons, those of us interested in both the possibilities of useful scholarship and the well-being of our students will continue to push ahead to try to invent what someday might offer our schools, our parents, our teachers, and our children more informative ways of representing what the experience of schooling means in their lives.

NOTES

1. For further discussion of the syntactical structure of forms of representation see Eisner, E. W. (1994). Cognition and Representation Reconsidered. NY: Teachers College Press.
2. This distinction is John Dewey's. See Dewey, J. (1934). *Art as experience*. NY: Minton Balch.

REFERENCES

Eisner, E. W. (1998). *The enlightened eye*. NY: Prentice Hall.
Kozol, J. (1991). *Savage inequalities*. NY: Trumpet Club.
Kotlowitz, A. (1991). *There are no children here*. NY: Doubleday.
Wiseman, F. (1980). *High school* [Film]. (Available from Zipporah Films, Cambridge, MA)

3. REPRESENTING THE TRUTH ABOUT PROGRAM QUALITY OR THE TRUTH ABOUT REPRESENTING PROGRAM QUALITY[1]

Linda Mabry

INTRODUCTION

> She spawned England's empire. . . . She was a larger-than-life royal with a genius for rule, who came to embody England as had few before her. . . . Determinedly molding herself into the image of a mighty prince, she made of England a true and mighty nation (McGeary, 1999, p. 165).

> Elizabeth's policy in Ireland, like that of Henry VIII, was based on two main principles: complete control of the whole country by means of government officials operating from Dublin, and complete uniformity in religion. This meant constant warfare, since the Irish were prepared to fight to the death for their freedom and for their Catholic faith (Ryan, 1966, p. 312).

Can both of these accounts of Elizabeth I's foreign policy be true? Or is neither true, both too superficial or too one-sided? Do these accounts turn a real person into a legendary character? Are they meta-narratives to which postmodernists would object because they deny Elizabeth, who lived out the history, her own view of her place in it – because they are historians' truth, not Elizabeth's truth?

Visions of Quality, Volume 7, pages 19–27.

ISBN: 0-7623-0771-4

How should history represent the English in Ireland – as bestowing order within and sometimes beyond the Pale, where the contentious wild Irish engaged in interminable local skirmishes, or as plundering so ruthlessly as to export Irish foodstuffs when a quarter of the native population starved to death in the infamous potato famine? How might we evaluate Elizabeth I's program in Ireland? What is the truth about this program?

TRUTH

All representation involves misrepresentation. Telling the truth is not a problem limited to nationalistic propaganda – Irish versus English views of a shared past, or Catholic and Protestant views of the modern troubles in six counties in Ulster. Historians as well as partisans disagree so deeply and frequently that history has been branded fiction. Can evaluators do any better?

How can an evaluator know the truth when truth wears many faces? All our sources of information are compromised. A program director says one thing, while personnel and stakeholders tell entirely different stories. Their expectations and experiences of the program differ. The problem is not just their unintended misstatements or evaluators' failures to listen carefully. Some clients promote their programs with effusive hyperbole, while their detractors and competitors exaggerate the missteps and fail to mention the achievements. How can evaluators be expected to recognize truth so elusive and so disguised?

Should we trust the truth (or, rather, the truths) of insiders – program directors or stakeholders – invested as they are? Or, the truth of the outsiders – are evaluators as disinterested as we seem? Is the truth of the evaluator more trustworthy because it is external, less obviously driven by advocacy (see Mabry, 1997a; Scriven, 1997; Stake, 1997), or does the lack of investment deny the evaluator critical purchase on the evaluand (Schwandt, 1994)? Decker Walker (1997) observed painfully that, because of their positions vis-à-vis the program and their particular responsibilities to it, clients and evaluators may be incapable of agreeing about the truth of a program's quality. The eye of the professional critic sees what the eye of the zealous implementer cannot see and vice versa.

The truth of the program director is not the truth of the stakeholder and is not the truth of the evaluator. The evaluator will have warrants for his or her truth, evidence and reasons. But so will the program director and the stakeholder. In a constructivist age in which everyone is constructing individual understandings, truth must be understood as idiosyncratic. We inexorably grind toward recognition that we cannot know what is true.

We are told we have erred in translating Einstein's relativity in physics into relativism in philosophy (Johnson, 1992; Rosenblatt, 1999), that the difficulties in understanding how time is affected by speed and how mass is related to energy are not at all the same as the difficulties in understanding what is true and what is right. We're stuck with uncertainty about what and whom to believe. Einstein's truth – or was it Irishman George FitzGerald's?[2] – has been superseded by quantum physics, competing views of scientific truth that may yet be reconciled in string theory (Hawking, 1999; Nash, 1999). Truth is fluid as well as personal, and no concentration of genius assures us. The hard sciences are as unreliable as the soft.

In matters of faith and morals, too, authority is questioned. Even Catholics believe the pope is fallible. We have revoked the monopoly on truth we once granted to religion. We have learned there are different religions, and they tell different truths. Christianity's Adam and Eve are as easily spurned as the Suquamish tribe's raven, discoverer of the first human beings inside a clamshell. We are no less suspicious of the monopoly on truth we granted to science, with its version of big bangs and odd mutations.

Evaluators cannot claim greater authority or more truth than Einstein, or the pope, or Chief Seattle of the Suquamish. We are defenseless against the onslaught of the postmodern crisis of legitimation (Roseneau, 1992; Wakefield, 1990).

REPRESENTATION

But shouldn't we tell the truth – at least, the truth as we know it? As evaluators, we claim the authority and expertise to represent programs. But how good are we at representing anything, even ourselves? Our lives are voyages of self-discovery, even (perhaps especially) for the most introspective of us. We cannot fail to doubt our capacity to understand and to represent program quality, to describe it adequately or accurately, to characterize anything outside ourselves. Others will have views as complex and personal as our own. How can we know a program well enough to represent it with the accuracy our *Standards* (Joint Committee on Standards for Educational Evaluation, 1994) demand?

If we must bring ourselves to recognize mystery in truth, we must also recognize that our descriptions of programs misrepresent. Are our reports examples of the distorted, disenfranchising meta-narratives postmodernists reject? Do we not turn people into mere stakeholders (Mabry, 1997b, 1998a, b)? We tell *our* side. Even when we try to tell *their* side, it's from *our* perspective. Should an objecting client's alternative representation be believed on the basis of intimate knowledge of the program, the deep understanding of lived experience – or disbelieved because of vested interest, defensiveness to protect professional prestige? If Elizabeth I had written an autobiography, would we believe her representation of sixteenth century English activities in Ireland? Do we believe Al Gore's history of the Internet?[3]

Disagreement between evaluators and clients is not news. But woe to the evaluator whose representation is unpalatable to a client willing to challenge its truth. Michael Scriven once described us as open to the most savage criticism, which is:

> You just didn't understand the other point of view . . . how can you possibly evaluate it properly. . . . This is where you lose the friends, [t]he people who aren't able to separate their ego from their work (Gierl & Migotsky, 1996, pp. 5–6).

An evaluator may consider a client's view of the program too bright, a client's view of the evaluation too dark, without being unable to adjust the contrast for maximum visibility by the viewing audience.

Given the tangled web of misrepresentations we work from and the misrepresentations we produce, would it be more or less useful for us to think of truth and fiction as indistinguishable? The word *fiction* is derived from the Latin *fictio*, meaning to fashion or to form. We fashion evaluation reports, but does this reduce them to fiction? Does it elevate them to fiction? Was Virginia Woolf right? She wrote of the task of describing the city of London:

> Only those who have little need of the truth, and no respect for it – the poets and the novelists – can be trusted to do it, for this is one of the cases where truth does not exist. Nothing exists. The whole thing is a miasma – a mirage (1946, p. 192).

Evaluation is a craft (Smith, 1997). Our reports are crafted. How crafty are we?

EVALUATION OF PROGRAM QUALITY

Our enterprise is to evaluate the quality of programs. But what is quality, and how may we detect it?

The Nature of Quality

Let us consider whether quality is a property of the program or an interpretation of the evaluator (or other beholders), whether quality is a construction of the program designers and implementers or of the evaluator. *If quality inheres in the program* (i.e. if quality is ontological), it is a reality to be discovered. Depending upon his or her acuity, the evaluator may or may not recognize it. It exists independent of the evaluator's recognition. *When judgments of program quality vary, they do so because of differences in evaluators' capacity to recognize it.*

Conversely, *if quality is an interpretation*, it is a creation of the evaluator (or other beholders). Not the program itself (or its designers and implementers) but the evaluator constructs an interpretation of the quality of the program. Interpretations of quality are based on acquaintance with the program, considered in the light of values and, perhaps, criteria and standards. Quality does not exist independently. The program exists, but its quality is a judgment about it. *When judgments of program quality vary, they do so because of differences in evaluators' capacity to construct it.*

Quality as a matter of perception or quality as a matter of interpretation – what matter? The distinction may not be important if we cannot confidently separate perception from interpretation. And we cannot. As we notice that the auditorium is crowded, we interpret whether that is a detriment to hearing of the listeners or a positive sign of program appeal. The interpretation forms while our sense of the size of the crowd forms.

Whether we think of program quality as inherent or constructed, our reports of program quality will reflect, to some extent, our own values and experiences. If quality inheres in the program, an evaluator's capacity to perceive it will be enhanced or obstructed by personal values and experiences. If quality is an interpretation, the evaluator's reckoning of its goodness will be influenced by personal values and experiences. Either way, our advocacy for our own values is inescapable (Mabry, 1997a) and our truths suspect.

Detecting Quality

Accurate representations of program quality depend on good design, good data, good analysis. Is good methodology enough? What is good methodology?

Different approaches can yield dramatically divergent judgments of program quality. For a volume illustrating a variety of evaluation approaches, Scriven and Elliot Eisner, among others, produced not only differentiated designs for an evaluation of a middle school humanities program but also starkly dissimilar findings of program quality (Brandt, 1981). Dissimilar? How could either hero be wrong? But there were lots of differences in methods and judgments of program quality in that publication. All our evaluation heroes have carved unique niches for themselves. The CIPP model developed by Dan Stufflebeam (1983) is readily distinguished from the evaluation approach favored by Egon Guba and Yvonna Lincoln (1989). Scriven (1994) and Bob Stake (1997) have insisted on contrasting methods of analyzing evaluation data. What assurance can there be when, along numerous professional dimensions, many respected colleagues contribute to an unfinished collective collage of resonating, overlapping, contrasting, contradictory positions?

Which question should we be asking: What is good methodology? Or, what good is methodology? Even if we could agree about methods, representing the truth about program quality requires more than sound designs and data collection, more than credible analytic methods. How are we to decide which issues should be focal, whether there are standards appropriate for determining program quality, which stakeholder interests should be prioritized, which groups constitute right-to-know audiences?

Consensus is comforting, but truth is not a matter of majority rule. Agreement is agreeable within professional communities and within each evaluation study. Trying to discern truth in contexts of dissensus is trying. But evaluation methods that reduce dissensus may merely suppress it, investing authority in evaluators and excluding others to greater or lesser extent. We have time-honored approaches, which tend to do just that: connoisseurship (see Eisner, 1991), expertise-oriented evaluation, accreditation, and blue ribbon panels (see, e.g. Madaus et al., 1987; Worthen et al., 1997).

Some latter-day approaches announce their determination to take seriously the values and perspectives of stakeholders and to build consensus – at least, working consensus. These approaches include participatory evaluation (Greene, 1997), empowerment evaluation (Fetterman, 1996), and utilization-focused evaluation (Patton, 1997). But stakeholders may disagree with the evaluator, or they may disagree irresolutely among themselves. When authority for conducting evaluation is dispersed beyond reclamation and when divisiveness becomes entrenched, the evaluation of a program can be thwarted. How *then* may quality be apprehended? Who is served by such an absence of authority?

But dare we reinforce the idea that there is *a* truth, *a* reality about a program, *a* proper way to evaluate it, *a* rightful evaluation authority?

IRRESOLUTION

The truth about program quality is occluded, making representation faulty. Our capacity to recognize and to understand truth cannot be perfected. Even if it could be, the impoverishment of all our languages and sign systems and communicative efforts would limit the quality of our representations.

Is evaluation hopeless? Well, yes. When does an evaluation report misrepresent a program? Always. There is always a different narrative that could be told, more emphasis given to the client's perspective or to the experiences of program personnel or beneficiaries or funders or competitors. Clients are right when they claim to have been misrepresented.

Even if quality is real, not merely in the eye of the beholder, our representations of program quality are our constructions. Evaluators are trying to help, trying to understand, trying to help others understand through our representations, but constructivists are right when they say everyone must inevitably have his or her own constructs. Postmodernists are right when they declare there are no universal criteria for adjudicating which representation is true. Evaluators' representations of program quality are wrong. But others' representations are also wrong. Do we need meta-evaluation to find or confirm the truth-value of conflicting claims? But who can confirm the truth-value of meta-evaluation findings (Mabry, 1998b)?

Recognizing the limitations of our representations can actually increase understanding. Dissatisfaction can spur our efforts to improve. And recognizing the limitations of our representations can lead us to proper humility in presenting conclusions, greater care about qualifications, less unwarranted confidence in our theories and findings and recommendations.

Or, anyway, that's my representation.

NOTES

1. Similar ideas were presented in papers at the annual meetings of the American Evaluation Association, Orlando, FL, November, 1999; the American Educational Research Association in Montréal, Canada, April, 1999; and the Stake Symposium on Educational Evaluation, University of Illinois, May, 1998.

2. "The Irish physicist George FitzGerald and the Dutch physicist Hendrik Lorentz were the first to suggest that bodies moving through the ether would contract and that clocks would slow. . . . But it was a young clerk named Albert Einstein, working in the Swiss Patent Office in Bern, who . . . solved the speed-of-light problem" (Hawking, 1999, p. 67). Whose truth *is* relativity?

3. Gore claimed, "I took the initiative in creating the Internet" (source: interview with Wolf Blitzer, CNN, March 9, 1999).

REFERENCES

Brandt, R. S. (Ed.) (1981). *Applied strategies for curriculum evaluation.* Alexandria, VA: ASCD.

Eisner, E. W. (1991). *The enlightened eye: Qualitative inquiry and the enhancement of educational practice.* NY: Macmillan.

Fetterman, D. M. (1996). *Empowerment evaluation: Knowledge and tools for self-assessment and accountability.* Thousand Oaks, CA: Sage.

Gierl, M. J., & Migotsky, C. P. (1996, April). *Educating evaluators: The untold story.* Paper presented at the annual meeting of the American Educational Research Association, New York.

Greene, J. G. (1997). Participatory evaluation. In: L. Mabry (Ed.), *Advances in Program Evaluation: Vol. 3. Evaluation and the Post-Modern Dilemma.* Greenwich, CT: JAI Press.

Guba, E. G., & Lincoln, Y. S. (1989). *Fourth generation evaluation.* Thousand Oaks, CA: Sage.

Hawking, S. (1999, December 31). A brief history of relativity. *TIME, 154*(27), 67–70, 79–81.

Johnson, P. (1992). *Modern times: The world from the twenties to the nineties*. New York: Harper Perennial Library.

Joint Committee on Standards for Educational Evaluation (1994). *The program evaluation standards: How to assess evaluations of educational programs* (2nd ed.). Thousand Oaks, CA: Sage.

Mabry, L. (1997a). Implicit and explicit advocacy in postmodern evaluation. In: L. Mabry (Ed.), *Advances in Program Evaluation: Vol. 3. Evaluation and the Post-Modern Dilemma* (pp. 191–203). Greenwich, CT: JAI Press.

Mabry, L. (1997b). A postmodern text on postmodernism? In: L. Mabry (Ed.), *Advances in Program Evaluation: Vol. 3. Evaluation and the Post-Modern Dilemma* (pp. 1–19). Greenwich, CT: JAI Press.

Mabry, L. (1998a). *Assessing, evaluating, knowing*. Proceedings of the Stake Symposium. Champaign-Urbana, IL: University of Illinois.

Mabry, L. (1998b, April). Meta-evaluation from a postmodern perspective (discussant). With Stufflebeam, D., Horn, J. & Nevo, D., *Metaevaluation: Methodology and recent applications*. Symposium at the American Educational Research Association annual meeting, San Diego, CA.

Mabry, L. (1999, April). On representation. In: R. Stake (Chair), *Representing Quality*. Symposium conducted at the annual meeting of the American Educational Research Association, Montreal.

Mabry, L. (1999, November). *Truth and narrative representation*. Paper presented at the meeting of the American Evaluation Association, Orlando, FL.

Madaus, G. F., Scriven, M. S., & Stufflebeam, D. L. (Eds). (1987). *Evaluation models: Viewpoints on educational and human services evaluation*. Boston: Kluwer-Nijhoff.

McGeary, J. (1999, December 31). Queen Elizabeth I. *TIME*, *154*(27), 162–165.

Nash, J. M. (1999, December 31). Unfinished symphony. *TIME*, *154*(27), 83, 86–87.

Patton, M. Q. (1997). Utilization-focused evaluation (3rd ed.). Thousand Oaks, CA: Sage.

Rosenblatt, R. (1999, December 31). The age of Einstein. *TIME*, *154*(27), pp. 90, 95.

Roseneau, P. R. (1992). *Postmodernism and the social sciences: Insights, inroads, and intrusions*. Princeton, NJ: Princeton University Press.

Ryan, S. J. (1966). Ireland: History to 1601 A. D. In: *Encyclopedia Americana, International Edition*, (Vol. 15, pp. 310–313). New York: Americana Corporation.

Schwandt, T. (1994). Constructivist, interpretivist approaches to human inquiry. In: N. K. Denzin & Y. S. Lincoln (Eds), *Handbook of Qualitative Research* (pp. 118–137). Thousand Oaks, CA: Sage.

Scriven, M. (1994). The final synthesis. *Evaluation Practice*, *15*(3), 367–382.

Scriven, M. (1997). Truth and objectivity in evaluation. In: E. Chelimsky & W. R. Shadish (Eds), *Evaluation for the 21st Century: A Handbook* (pp. 477–500). Thousand Oaks, CA: Sage.

Smith, N. (1997, March). Professional reasons for rejecting a handsome evaluation contract. Symposium paper presented at the annual meeting of the American Educational Research Association, Chicago, IL.

Stake, R. E. (1997). Advocacy in evaluation: A necessary evil? In E. Chelimsky & W. R. Shadish (Eds), *Evaluation for the 21st Century: A Handbook* (pp. 470–476). Thousand Oaks, CA: Sage.

Stake, R., Migotsky, C., Davis, R., Cisneros, E., DePaul, G., Dunbar Jr., C., Farmer, R., Feltovich, J., Johnson, E., Williams, B., Zurita, M., & Chaves, I. (1997). The evolving synthesis of program value. *Evaluation Practice*, *18*(2), 89–103.

Stufflebeam, D. L. (1983). The CIPP model for program evaluation. In G. F. Madaus, M. S. Scriven & Stufflebeam, D. L. (Eds), *Evaluation Models: Viewpoints on Educational and Human Services Evaluation* (pp. 117–141). Boston: Kluwer-Nijhoff.

Wakefield, N. (1990). *Postmodernism: The twilight of the real.* London: Pluto Press.

Walker, D. (1997). Why won't they listen? Reflections of a formative evaluator. In: L. Mabry (Ed.), *Advances in Program Evaluation: Vol. 3. Evaluation and the Post-Modern Dilemma* (pp. 121–137). Greenwich, CT: JAI Press.

Woolf, V. (1946). *Orlando.* New York: Penguin. (Original work published 1928)

Worthen, B. R., Sanders, J. R., & Fitzpatrick, J. L. (1997). *Program evaluation: Alternative approaches and practical guidelines* (2nd ed.). New York: Longman.

4. COMMUNICATING QUALITY

David Snow

INTRODUCTION

I have a small backpack that is full of quality. Adjustment straps and gear loops are tightly sewn into its sturdy red nylon fabric, it fits comfortably against my back with cushioned shoulder straps, and its thick belt distributes weight evenly under heavy loads. The hand-stitching and sturdy design of the pack by one small Colorado company has served me well without visible wear for several years. It is small enough to serve as a book bag for my schoolwork and as a knapsack for my travels, yet the pack has the capacity and rugged construction necessary for short wilderness trips. As I mentioned, this pack is full of quality.

Just after purchasing this pack a couple of summers ago, I stuffed it with an afternoon's gear and headed out with a group of family and friends to climb one of the smaller mountains in Colorado's primary range. By midday the altitude had encouraged the recreational climbers in our group to turn back while my brother and a friend remained to join me to the peak. Through most of the climb I was following a few paces behind my brother who was leading our trio, and I in turn was followed by my friend Mark at my heels. The walk would have been an easy few hours had there been air to breathe, but frequent, short rest stops were necessary to satisfy my thirsty Midwestern lungs.

As we climbed through the early afternoon we eventually emerged from the small trees and scrub at the tree line into the open tundra. The panoramic view provided a good excuse for another stop so we could all catch up with our need for air and have a good look around. As I turned to Mark he pointed to my shoulder strap and asked, "Is that a new backpack?" Apparently, while I had

Visions of Quality, Volume 7, pages 29–42.

ISBN: 0-7623-0771-4

been climbing behind my brother's heels and thinking about oxygen, Mark had spent at least some of his time examining my pack. I do not remember the exact words of the conversation that followed, but I am certain that his comment had me spouting some product specifications – capacity, material, construction, price – followed by a customer testimonial that would have made any company proud. I am also sure that I mentioned an old duffel bag of mine that was made by the same company which is still in great shape today after years of use and mild abuse. Come to think of it, I probably even added that I knew of several wilderness programs that swore by this company's products.

The rest of the climb was enjoyable. The intermittent conversation drifted away from the topic of my gear just as the beauty of the surroundings took my mind off of the thin air. We made it to the peak despite the wishes of the mountain goats that had been standing steadfast on the ridge, and made it back to our car by nightfall.

Later that same week the topic of my pack resurfaced. While making plans for a rainy afternoon, Mark let on that our mountain top conversation had held a place in his thoughts. He had been examining my pack more carefully, and wanted to take a trip to that same company store to have a good look at what they might have for him. Off we went. It seems that I had succeeded in conveying to him the quality of my pack.

COMMUNICATING QUALITY

I have trouble imagining a more perfect set of circumstances for conveying the quality of something to someone else. Mark and I had all the time in the world, our activity had him interested in learning more about my pack, I was certainly willing to share the information carefully, and we knew each other well enough for him to trust my judgment and well enough for me to know his preferences.

Since leaving that mountain, however, I have come to realize just how rarely I experience the careful communication of quality. Close, careful communications are few and far between. The loudest voices I hear seem to be from those with some vested interest such as advertisers, politicians, and entertainers. I even find that my "water cooler" conversations about movies and books are more central to understanding the people conversing than they are in communicating the quality of their topic. All too often I invest time or money in a product of less than acceptable quality. Where is that mountain top conversation when I really need it?

I suppose we are all in need of additional understandings that would result from conversations like the one I had with Mark. It is the nature of our society to accept a synthesis of quality that is provided for us by waiters, GAO officials, and the like, but careful attention to the effort of conveying quality is seldom the rule. To the extent that we all benefit from increased understanding of quality and seek to provide a sense of quality to others – sharing our tastes in food, expressing the performance or appeal of others, and maybe even forging careers in critical and evaluative studies – it would seem that an emphasis on the process is in order.

I also suppose that we have all come to understand at some level (especially the professional critics and evaluators) the challenge that confronts us as we embark on efforts to communicate quality. It is, after all, an understanding that is best reached first hand. I knew exactly how much I disliked the movie I saw the other night as soon as it was over. The thorough understanding that it did not appeal to me, however, came about an hour and a half too late. But, being so certain that the movie was one that I would enjoy, I understand the challenges faced by a critic who would have liked to have saved me the effort and expense.

I am using the phrase "communicating quality" here to emphasize the interaction that takes place in the best of these cases. Again, my mountain top conversation with Mark serves as a good example. After all, had he not shown interest in the pack's specifications and my other evidence of quality, I could have expounded instead on the pack's aesthetic appeal, or even suggested the importance of supporting local companies. On the mountain I had been watching Mark's expressions and reactions in order to find the best way to bring him to an understanding of the quality that I saw in this pack. As I have indicated, however, this is a rather extraordinary example of communicating quality. I suppose that another example might better illustrate the import that this idea deserves.

It may be more realistic to imagine a scenario where a veteran teacher is charged by her principal to evaluate the quality of a new teacher in her school. This veteran teacher is expected to convey a strong sense of this quality to her principal, and in so doing she must consider not only the quality that she observes, but also must attend to the delivery of that which is observed. This means that she will need to move beyond a mere representation and involve herself in an interaction with the principal that will encourage the principal's thorough understanding of the quality of the new teacher.

Expanding on this example, imagine that the veteran had accepted this charge – evaluating the performance of a new teacher at her school – and had decided to rely solely on a written report of her findings for her principal. Imagine further that in this report she had written about a group of students who "struggled to understand the reading that they had been assigned the night before." The veteran teacher, seeing the students struggling, was encouraged by the challenge presented by the new teacher. This observation was then followed by several specific examples of observed behaviors that describe a group of children productively grappling with new ideas that had been encountered in the reading. This section of the report is then read by the principal who imagines a teacher "struggling" to be effective rather than one who is engaging a group of students who are "struggling" through a challenging experience. When this image is carried through the rest of the report, despite the examples that supported an image to the contrary, the report has failed to communicate the quality that was observed by the veteran teacher.

There certainly do exist readers who could have used this veteran teacher's report to come to a good understanding of the quality that had been observed. So, the report itself is not particularly faulty. But even so, the report failed in the one thing that it was supposed to do. It failed to convey the observed quality to the principal. No statement of the "accuracy" of this report is valid in the light of the resulting miscommunication. In fact, this example illustrates the sort of misunderstanding that is likely to occur at some level whenever the reader is not taken into account. Terms such as "evaluative reports" and "representations of quality" give the false impression that a sense of quality can be passed along from one person to the next without concern for an interactive process necessary for communication, and it is this lack of concern that creates great potential for misunderstanding.

This veteran teacher made two mistakes. First, she chose not to deliver or at least follow-up on her report in a face-to-face conversation with the principal. If the veteran teacher had been in a conversation with the principal when she said that a group of students "struggled to understand the reading that they had been assigned the night before," the phrase would have elicited some reaction from the principal. Even if this reaction were a subtle one, it would have indicated to the veteran teacher a need for clarification. This clarification would have taken place before moving forward in the report, and therefore further misunderstanding would have been avoided. Here the conversation between these two professionals becomes the key to the communication of teaching quality.

The second mistake made by the veteran teacher is that she did not take into consideration the preferences, experience, and leadership style of her principal. Regardless of her ability to meet with the principal face-to-face, this teacher's familiarity with the principal should have been used to inform an approach to the report that she presented. Her awareness of the principal's lack of regard for constructivist theory, for example, might have influenced her to change the language of her report to exclude her use of the term "struggle" which in turn would have reduced the likelihood of the misunderstanding that resulted.

I do not want this call for face-to-face conversations in the communication of quality to be read as an argument for proximity to allow for interaction. Face-to-face meetings facilitate interaction, but these meetings will not ensure that an understanding of quality is conveyed. The important point here is that thorough communication of quality is dependent upon careful interaction. The give and take experienced by Mark and me on that mountain could easily be translated into a principal's office, or even into the report of a veteran teacher as long as she takes her reader into account. In a good report of quality direct interaction is replaced by an interaction between the reporting teacher and her knowledge of her principal. She may ask herself what her principal might want to know about this new teacher or what the principal is likely to think about the words she has used in her report. Having produced the report with these questions in mind, the principal will then interact with the report in a fashion that will ensure the understanding that is sought.

All of this is well and good, but the reality of evaluative reports of all sorts often forces us to consider an even more distant audience, an audience about which little is known and one that therefore makes the interaction that I have described difficult. Even the veteran teacher may have readers – board members, parents, a superintendent – whom she has not met. Her written report is her only means of communicating the quality she has observed, and she has little or no knowledge of the people to whom she is reporting. Is her effort hopeless, or is there something to be learned from her experience in trying to communicate with her principal? The question in a more general form is an important one for all evaluators to consider: How can those seeking to communicate quality best approach their work?

COMMUNICATING WITH THE DISTANT READER

From the beginning of my interaction with Mark on that mountain top, I was advantaged by the fact that I knew him. Mark and I had a common understanding of many things long before our discussion about my pack. A stranger would not have known Mark's appreciation for craftspersonship or his interest in versatile products. Communicating the quality of the pack was an exercise in relating to him as I had done so many times before that day. It was a communication of quality facilitated by my understanding of my audience and my knowledge of how best to proceed.

The reporting teacher in the second example, on the other hand, would not be expected to have the same level of familiarity with her principal. Despite this relative disadvantage, I still suggested that she should have used whatever existing knowledge she had to communicate her observations of teaching quality. The relative lack of familiarity with her reader may mean that she needed to work harder: considering her words more carefully and searching for ways to address the issues that she knows are important to her principal. If she had given careful attention to the effort, the resulting report would almost certainly have been a more effective representation of the quality she had observed.

As I have suggested, however, there is no reason to believe that evaluators should expect to know their readers on a personal level. This lack of personal knowledge presents the evaluator with an obstacle in the effort to communicate quality. A professional theatre critic, for example, does not have the time to discuss his review with each of his readers even though such interactions would enhance the communication that is his goal. He also does not know many of his readers personally and is therefore unable to speak to their individual preferences (even if his medium allowed for individualized reviews). His ability to interact, either directly or with his understanding of his readers, is marginalized by a lack of available familiarity.

But we will assume that this critic is conscientious and that he is driven to find a way to communicate the quality of the show he has just seen. He may ask himself whether or not to encourage his readers to attend a show, and he may also struggle with the possibility that the show may be a deep experience for some even though he found it to be shallow. How can he best convey the quality of the show he has experienced? I would say that the answer to this question is again rooted in the idea of a familiarity that enhances interaction.

When we peel away the direct personal experiences that give us familiarity with those we know, we are left with a familiarity bred by common experiences. I am not referring only to the common experiences of specific social, ethnic, or religious groups, but instead to a common experience that can be described as a general shared human experience: being caught in a traffic jam, sore muscles and joints that appear with age, and the joy and frustration of being a parent. It would seem that these common experiences can (and often do) provide a foundation of common understanding upon which effective communication can be based.

Ingrained in all of these experiences are the words that are used to recount them. These words are part of a common language and are heavily laden with a variety of sensations. The mention of a traffic jam on a hot afternoon brings to my mind the smell of hot pavement, the sticky discomfort of a clinging shirt, and the frustration of having to change afternoon plans. People might say that their pizza tasted like a cardboard box, or that their trip in a cab might as well have been a ride in a bumper car, or that they feel as though they have been up all night. These things are said by people working to convey a rich sense of that which has been experienced, and these people are communicating in a way that moves far beyond the words that they use.

The term "working" is an important word here as I do not see the effort as being an easy one. I would say that when an audience is as unfamiliar and disparate as the audience of a theatre critic, for example, the effort is a formidable one. The critic needs to be careful in his work because the power of common language to invoke a rich experience can work against him in much the same way as it works for him. (Remember that the principal was led by his experience to a misunderstanding after reading the word "struggle.") It is fairly clear that care must be taken, not only in the consideration of the audience, but also in the choice of language.

So, in order for our conscientious critic to communicate the quality that he has experienced, he must start with the use of this common language. He needs to find words, images, and descriptions that are familiar to his readers, and he needs to use them to build a review that will convey the quality he has observed. When this is accomplished, an interaction can take place between the critic's reader and the review itself. The reader will be exposed to the familiarity of common language, this language will then draw in past experiences, and the two will inspire a dialog that will carry the reader toward a greater understanding of the show's quality. Without the common language and the familiarity it provides, the reader is left outside the review and the experience that it could have provided. When the critic has succeeded in communicating quality, the reader will be able to make an informed decision about attending the show, and the critic will have accomplished what he set out to do.

These considerations indicate the difficulties encountered by the professional critic who approaches the job seriously. In fact, anyone choosing to communicate the quality of something as complex as a staged production is going to be faced with a number of difficult choices. For example, should the critic share personal preferences with the reader? Having a sense of the critic's taste would surely be helpful in determining whether or not to take the review at face value. Or is the alternative the more effective approach? Should theatre critique be strictly an effort to carefully describe the show in purely objective terms? In a related concern, a critic would also need to consider whether or not to explicitly recommend the show. Should the critic rate the show, or so severely praise or pan it that a recommendation of one sort or another would be obvious? The alternative would be to carefully communicate the quality of the show without making this sort of prescriptive statement. Without a judgment the critic would be helping the readers to understand the show's quality and then allowing them to make their own decisions. These concerns begin to suggest the depth of complexity inherent in communicating quality to a distant reader.

INTERACTING WITH FLAMINGO

I understand that it is one matter to suggest how a communication of quality should be produced and quite another to take responsibility for the effort itself. My charge of producing a report that interacts with a distant audience, for example, carries with it a wealth of difficult choices. To come to a better understanding of these difficulties, and to possibly contextualize the ideas that I have been pursuing thus far, I chose to produce a modest critique of my own. The following is an effort to draw my readers into a conversation with my understanding of one sculpture's quality.

Flamingo (Alexander Calder, 1898–1976) has become a Chicago landmark. This early product of the federal Art-in-Architecture Program has come to symbolize the unique visual qualities of the city that surrounds it. The tall red arches of the steel sculpture complement the black and gray buildings lining the streets that define the city's Federal Center and its Plaza.

At first glance the work has a strange but attractive appeal. The shape is an inconsistent mixture of straight lines, curved edges, and sharp corners, but each of the long arches and broad legs taper to narrow feet that appear to be poking into the plaza pavement as if the work had been dropped and left stuck in its place. An imaginative viewer may see a long-legged spider, or the sweeping skeleton of some mythical creature, or even a great fountain of cold red steel in this forty-foot structure. For most, however, the first impression will merely call the viewer in for a closer look at Chicago's Calder.

The seemingly random planes and irregular shapes of the piece give it an unpredictability that the viewer is likely to find mildly confusing. I found myself walking from one side of the work to the other trying to anticipate how my next perspective would look, but I was wholly unsuccessful. Each view brought with it a new experience. The plates between the short legs of the work would disappear when they were viewed from their edges, and then expand to shift the balance of the work when viewed at their faces. The large arches would draw focus and then fade to insignificance as they retreated from the foreground. The frustration of predicting the work soon gave way to enjoying the ever-changing form of Flamingo as I moved around the plaza.

The changing profiles of the sculpture suggest a subtle but satisfying motion to the moving viewer. Calder is known for his work with hanging mobiles and other dynamic forms, but Flamingo offers the viewer a chance to control the motion that is experienced. Even the commuters who pass through the plaza each day are exposed to a dynamic presentation of Calder's work (although it would be difficult to say how many of them notice it).

This work is most engaging in its ability to challenge the viewer to understand it. The seemingly simple shapes of its parts and its uniform color suggest a predictability that soon gives way to the complexity that I have described. Whether the viewer is caught up in trying to predict the work's profile, as I was, or is drawn to its subtle irregularities, the memory of it all fades quickly as the eyes are drawn away. You will know this recognizable piece when you see it again, but you will never remember it fully.

It is my hope that I have communicated some sense of the quality of Calder's Flamingo to my readers. Interestingly, I found the process of refining the final draft to be somewhat frustrating because my personal experience with the sculpture made me entirely unqualified to judge the quality of my representations. Once the work was studied there was no way for me to return to the lack of familiarity that most of my readers would likely share. This problem seems largely unavoidable, but reinforces the need for external readers prior to general consumption. Understandably, this concern is clearly addressed in published works by the editors and reviewers of newspapers, magazines, and journals.

It is also interesting to me that as I wrote this critique I was forced to surrender my initial objectivity in favor of a number of subjective comments. This subjectivity was demanded by what I deemed to be a more effective communication of my personal experience. In discussing the changing profiles of the piece, for example, I thought it best to describe the mechanism through which I reached the observations made. I wanted the reader to see me walk, and to walk with me, so that we would experience together the changing faces of the work. Without building (or at least trying to build) this common experience I would not have a foundation on which to build an understanding of what I see as Calder's expressed quality of motion.

In fact, an ebb and flow of objectivity and subjectivity seems to carry through this critique. I was clearly sharing my personal sense of the work throughout this attempt at an objective review with my emphasis on descriptive form and use of fantastic imagery. In retrospect, I see that it would be impossible to communicate to a distant reader any level of quality without injecting into that communication some of my personal sense of the work. I believe that an effort to maintain objectivity would have hampered my ability to best communicate the quality that I had seen in this sculpture.

The alternative to this objective and subjective balance would be a dominant form of one or the other approaches. It seems to me that a largely objective representation of the work would have yielded an impersonal and potentially dry critique. Such a review might include these words: "The greatest of the arches rises 40 feet above the Federal Central Plaza. Its red steel is dotted with the fastened bolts that cover the large plates in the core of the work." The words bring the reader closer to the work, but ignore the potential for reader interaction. In contrast, a subjective approach would inject personality, but would lack the substance that objectivity provides: "Calder's Flamingo is simply beautiful! I have never enjoyed myself more than I did the afternoon that I spent in the Federal Center Plaza."

As ridiculous as this purely subjective quote may sound, this sort of approach might have served me well in my mountain top conversation with Mark. Since we already knew each others' interests and preferences, it seems that there would have been little need for explanations of my reasons for liking my pack. This observation leads me to believe that objectivity begins to play a greater role in the effectiveness of communicating quality as personal familiarity is removed from the effort. The foremost concern in the mind of any critic, therefore, should be the use of objectivity and subjectivity in a combination that best serves the ultimate goal of communicating quality.

Prior to presenting the critique of Flamingo I had also suggested that a conscientious critic would need to choose between a prescriptive or descriptive tack. In the case of my critique, I avoided making explicit judgments of the quality of Calder's work. Since I had not given the reader a thorough sense of my tastes in art, I supposed it unlikely that my rating of this piece would carry much meaning. A professional critic may be accused of shirking his duty if he were to present a review of this type, but my academic standing carries with it no such political pressures. I did, however, let my feelings be known in more subtle ways. Phrases such as "unique visual qualities" and "attractive appeal" indicate my appreciation of Calder's Flamingo, and would have been hard to avoid. Here too, it becomes clear that a subtle balance between prescriptive and descriptive choices should be a function of a desire to communicate quality most effectively. Had I more carefully avoided sharing my preferences, or dwelled upon them for that matter, any communication of quality that I was able to achieve would likely have suffered.

COMMUNICATING PROGRAM QUALITY

What I have avoided to this point is any discussion that involves communicating a relatively complex quality. Specifically, none of the thoughts presented here would be of much worth if I did not consider the implications that they have on the salient issues of program evaluation. Like the complexity of the programs themselves, the complexity of an effort to evaluate a program is enormous. A great number of dynamic elements come together in programs – activities, services, staff, goals, communities, and politics. A public school has many classrooms, a social services agency has a number of caseworkers, and a community center has many smaller programs. Each part is one small piece of the whole, each part has its own quality, and each part contributes differently to the quality of the whole. It is not often clear how all of these elements interact in a complex program, but it is clear that a full understanding and communication of the quality present is a difficult job.

This complexity, when fueled by the social importance of these programs, brings out differing opinions as to the nature of program quality. What is good teaching? How many cases can a social worker effectively handle at once? What is the most 'effective' effort? These are difficult questions, and they are not questions that are going to be answered by consensus. This lack of agreement not only adds to the complexity that we may seek to evaluate, but it also carries with it a lack of common language with regard to these programs. Although such disagreement is natural and even healthy for the programs themselves, it does decrease the likelihood of communicating program quality effectively.

If we agree that it is actually the quality of a program that needs to be represented (that a qualitative approach is needed), then the previous examples and discussion indicate a direction for program evaluators. This is not to suggest that quantitative methods should be ignored, but rather that these and other objective information should be employed in a greater effort to communicate the quality of the program in question. Quite simply, I believe I have indicated that conveying this quality depends on the evaluator's careful and thorough attention to an audience interaction with the program's quality.

Given the fact that program evaluators are often not familiar or even able to interact personally with their audiences, the ability of the written report to interact with its audience becomes essential. Rather than simply accepting the prevalence of written reports of this type, it seems to me that the program evaluator should hesitate for a moment and recognize the compromise that must be made in cases where direct interaction is not possible. The evaluation report is to be the common thread between the experience provided by the program and the report's audience, which is a great burden for its creator to bear. And insofar as the program experience is a rich and complex one for the evaluator to communicate to her audience through the report, the difficulty of the task is amplified.

IMPLICATIONS FOR THE PROGRAM EVALUATOR

Before I enter into a distillation of my thoughts with regard to program evaluation, I feel a need to justify my focus. I have been referring primarily to a communication and potential miscommunications between the evaluator and reader, but I have also alluded to other interactions. I suppose there is actually a chain of communication to be considered here: program to evaluator, evaluator to report, report to reader, and even reader to politician or funder. But, due to what I see as the potential for the report to be the weak link in this chain, it has drawn my attention as the focus of my discussion. It is with this in mind that I offer the following suggestions.

First, the evaluator should try to ensure interaction by consciously developing familiarity with the reader. Such familiarity is produced as a byproduct of any written account (a point that I believe is illustrated by my descriptions of Flamingo and my backpack), but the evaluator may also need to consider a more explicit approach to self-revelation that will provide the familiarity needed to encourage interaction with the report's readers. As I noted in my discussion of my critique of Flamingo, familiarity serves to support subjective information that gives life to reader interaction.

Second, there may also be a need to compensate for some lack of familiarity that will always remain in these cases. Lack of familiarity can be compensated by objectivity, and this idea combined with the previous suggestion directs the evaluator to achieve balance in her approach. I am referring to this sort of tenuous balance when I call for care in creating evaluation reports.

Third, the evaluator should keep in mind the importance of the report's ability to induce interaction. Beyond concerns regarding familiarity, common language must be sought and used to ensure a thorough understanding of the program that was observed. The common language calls the readers' past experiences to mind and uses them to create a vicarious experience of the program at hand. An understanding of a report written about a second grade classroom, for example, is enhanced by the personal experiences that we have all had as visitors, parents, or, at the very least, second graders. The evaluator responsible for communicating the quality of this classroom would be remiss if she were to avoid helping her readers to remember what it was like to be back in a second grade classroom: the smell of Playdough, the little chairs, the sound of a pencil landing on the tile floor. A successful evaluator pushes her readers beyond the written words and into the quality of the program.[1]

The program evaluator should note, however, that common language used to enhance the communication of quality also has the potential to cause miscommunication. Language thought to be common may inspire an experience in opposition to that which was intended. This is a very real concern when an evaluator becomes too familiar with the program to synthesize the quality of it successfully, and is one more good reason to take advantage of the process of metaevaluation. Through this process a thoughtful discussion with a third party may replace the strong need for direct interaction with the audience of the report. Metaevaluation would bring to light elements of a report that were potentially misleading and enhance the communication of quality offered by the report.

Last, I will again suggest a need for program evaluators to maintain a goal of effective interaction. Every available tool and approach should be used with careful attention given to an overall communication of the quality that they have experienced.

SUMMARY

When I hear the term "report" or "representation" applied to the concept of expressing quality I feel as though I am expected to believe that an understanding of quality can be delivered in a nice, neat bundle. Granted, the delivery of information – numbers, dimensions, effects – can be an important part of such an expression, but it seems to me that the quality resides in and among these descriptors. By its very nature, therefore, quality is difficult to "report." The only way to express this quality is through a concerted and careful effort of communication. It is for this reason that I prefer to limit my use of the term "reporting" to expressions of quantity, and my colleagues will hear me referring to the "communication" of quality.

As I have noted, I see the communication of quality is an interactive process, whether this interaction takes the form of two friends talking about the quality of a backpack, an evaluator discussing the quality of a classroom teacher, or a critic's review speaking to its readers. In any case the effectiveness of the process is dependent on the interaction that takes place in the mind of the person who is accepting a representation (a *re-presentation*) of quality. The communicator's careful use of familiarity or some common language encourages this interaction and therefore enhances the communication of quality.

I also used this forum to suggest that the complexities and responsibilities of social programs bring great importance to the effort of communicating quality. Given this importance, I recommend that program evaluators use descriptive and prescriptive methods, as well as subjectivity and objectivity, as tools to extend the capability of their work to communicate the quality that has been experienced. Again, their ability to communicate this quality rests upon the interaction that takes place between evaluator and audience. As I see it, the job of every evaluator, reviewer, and critic is to attend carefully to what has been described here as the communication of quality.

NOTE

1. I seek only to add an emphasis – one of interaction – to a concept that has been presented previously. Stake and Trumbull (1982), for example, wrote of vicarious experience in their description of what they termed "naturalistic generalizations." These authors suggest that careful representations can take the place of personal experience in conveying quality. "Naturalistic generalizations are conclusions arrived at through personal engagement in life's affairs or by vicarious experience so well constructed that the person feels as if it happened to themselves" (Stake, 1995, p. 85). Eisner (1998), too, attends to the important concept of careful representation: "Humans have a basic need to externalize the internal, to communicate, to share their experience with others. The trick is to learn how to use a form of representation through which imagination, affect, and belief can be given a public, articulate presence" (p. 235).

REFERENCES

Eisner, E. (1998). *The enlightened eye: Qualitative inquiry and the enhancement of educational practice*. Upper Saddle River, NJ: Merill.

Stake, R. (1995). *The art of case study research*. Thousand Oaks, CA: Sage.

Stake, R., & Trumbull, D. (1982). Naturalistic generalizations. *Review Journal of Philosophy and Social Science*, *7*(1), 1–12.

UNDERSTANDING PROGRAM QUALITY: A DIALECTIC PROCESS

Claire Lloyd

ENCOUNTERING QUALITY

One of the principal tasks confronting the evaluator is to seek an understanding of the properties of an evaluand. There are many properties that may be assessed, but evaluators typically make a "fundamental and important" distinction between merit and worth (Scriven, 1993, p. 67). Merit concerns the goodness or effectiveness inherent within an evaluand. The evaluator focuses upon a program's internal functioning to assess such things as the extent to which the client's needs have been met, the effectiveness of the program's organizational structure and implementation procedures, the relationship between program inputs and outcomes, and so on. Worth refers to the value or efficiency of an evaluand. The evaluator looks beyond a program's inherent goodness to assess its broader impact in terms of cost, risk, or some other dimension (Scriven, 1993; Shadish, Cook & Leviton, 1995). Recently, however, Robert Stake and Elliot Eisner have brought another important evaluative property to the fore: program quality. Stake writes: "A basic content question for all program evaluation is: What is the quality of this program? It is the question most directly derived from the meaning of the word 'evaluation' and is the feature that distinguishes evaluation from other research" (Stake, 1998, p. 10).

To date, quality has received little more than a passing consideration from many within the field. It is typically discussed in conjunction with merit and worth – though theorists position it differently in relation to these properties.

Visions of Quality, Volume 7, pages 43–54.
2001 by Elsevier Science Ltd.
ISBN: 0-7623-0771-4

For example, Michael Scriven (1993) uses merit and quality interchangeably when referring to the intrinsic goodness of an evaluand. Scriven writes: "The merit of teachers is a matter of how well they teach whatever it is that they teach ... merit means quality according to the standards of the profession" (p. 67). In contrast, Jennifer Greene (1997) distinguishes between quality, merit, and worth: "To evaluate is, according to tradition, to judge fairly the quality, merit, and worth of a program" (p. 26). Yet she does not discuss the relationship between these properties.

Stake and Eisner (1998) challenge both the complacency the field has shown toward quality and the perception of quality as yet another evaluable property that can be positioned within or alongside merit and worth. According to their way of thinking, quality refers to a generic sense of program conditions – one that acknowledges the complex socio-political system within which programs are situated and, in turn, act upon (Stake & Eisner, 1998). The assessment of program quality requires two evaluative tasks: descriptions of program composition or texture and judgments of the multiple meanings of merit or worth. To describe program composition the evaluator must pursue an understanding of the complex contingencies that inform stakeholders' perceptions of the program and shape their program experiences. These descriptive accounts represent "the striving, the context, the sense of episode, challenge, and the multiple realities of stakeholders" (p. 1). To judge program quality the evaluator must convey the "multiple meanings [of] merit, well-being, worth, integrity, and goodness" (p. 1). Stake writes: "Goal attainment, productivity, success, effectiveness, comprehension of 'what works', all are part of the definition of quality but subordinate to it" (Stake, 1998, p. 10).

Stake and Eisner's definition of quality reflects its epistemological origins. Quality was conceptualized in response to the influence of postmodernism on evaluation in light of a concern that "good evaluation within a modernist reality may not be good enough in postmodern contemplation" (Stake, 1998, p. 1). Thus when seeking to understand program quality, postmodern assumptions shape perceptions of the evaluand and, consequently, define the nature of the evaluative process.

Postmodern Assumptions

Postmodernism has emerged from a dissatisfaction with the fundamental assumptions of modernist epistemology and Enlightenment rationality (Burbules, 1995; Kincheloe & McLaren, 1998). At least three broad insights have contributed to this state of affairs. First, the emancipatory ideals of modernist epistemology fail to recognize the way institutional and informal power relations impact society as a whole and shape the everyday experience of individuals and groups. Second, traditional interpretations, which focus on unity, ignore the diversity both within and between groups along dimensions of race, class, and gender. Third, traditional "universal" representations of reality are discursively situated in the social and material conditions of the context in which they are constructed (Burbules, 1995).

Yet although postmodernism casts doubt upon traditional assumptions and representations of reality, it is not considered an alternative; instead, it is typically described as a mood or attitude characterized by skepticism and uncertainty (Burbules, 1995; Kincheloe & McLaren, 1998; Stake, 1998). Postmodernists challenge the legitimacy of the knowledge frameworks generated by traditional approaches. Moreover, in light of the social and institutional structures, both locally and in larger systems, that impact individuals and groups in different and unpredictable ways, postmodernists are mindful of the complexity of the inquiry process and the fallibility of the investigator's efforts to make sense of diverse knowledge constructions (Burbules, 1995).

Epistemologically, postmodernism is most consistent with constructivism. Constructivists argue that reality is dependent for its structure and content on the efforts of individuals and groups to make sense, or construct an understanding, of their experiences (Guba & Lincoln, 1998). Hence, multiple constructions of reality are assumed to exist which are considered "more or less informed and/or sophisticated" (Guba & Lincoln, 1998, p. 206). For constructivists the inquiry itself is a social construction or, as Beck (1993) writes, "an interactive process of knowledge creation" (p. 4). The investigator and the individuals under investigation engage in a "hermeneutical and dialectical" exchange to compare and contrast their different constructions (Guba & Lincoln, 1998, p. 207).

Postmodernism departs from constructivism regarding the purpose of the inquiry. Whereas constructivists seek to synthesize different constructions into a "consensual and multi-voice reconstruction that is more informed and sophisticated" (Guba & Lincoln, 1998, p. 207), postmodernists question whether reality can be legitimately reflected in synthesis in light of the complex and diverse constructions surrounding an issue or concern. Stake (1998) writes: "We see things differently, we have different values. Consensus fails to address the realities of diversity in the society" (p. 7). For postmodernists the dialectic exchange is undertaken to reveal new and more complex understandings of the diverse constructions under investigation.

The Dialectic Model

Stake et al.'s (1997) dialectic model of evaluation demonstrates the way an evaluative process grounded in postmodern assumptions can be used to pursue an understanding of program quality. Stake and his associates introduced the model in response to Michael Scriven's (1994) paper "The Final Synthesis" which recommends the application of a rule-governed logic to facilitate valid evaluative conclusions. According to Scriven's logic, the evaluator performs a needs and wants assessment to identify a "comprehensive list of criteria and tentative weights" which are presented to the client for review (p. 377). Once appropriate criteria and standards have been established, performance is measured and the data synthesized into an overall judgment of merit or worth. Contrary to Scriven's approach, which seeks synthesis, the dialectic model "works toward the perspectival, the conditional, and comprehensive" (Stake et al., 1997, p. 97). Based on the notion of an evolving synthesis, understanding grows increasingly complex through a dialectic process of knowledge construction. The evaluator gains insight into program composition as stakeholders' program experiences are explored from "different points of view and different frames of reference" (p. 96). An awareness of the multiple meanings of quality is developed and refined throughout the dialectic process as the evaluator "collects the values of others, [and] skeptically scrutinizes them without insistence upon the superiority of [his/her] own judgments" (p. 97).

The excerpt below demonstrates how Stake's dialectic model was employed to pursue an understanding of the quality of an inservice training program.

PURSUING QUALITY

Planning the Pursuit

In the fall of 1998, the faculty in the Department of Special Education at the University of Illinois began the fifth cycle of a year-long inservice training program for professionals working in the field of supported employment. The program targeted professionals providing services to individuals with severe disabilities who wish to enter the workforce.

Program Description

Over the past decade, legislation has been passed upholding the rights of individuals with severe disabilities to participate fully in society. Moreover, federally funded demonstration projects have shown that if these individuals receive appropriate support they can engage in community activities and experience successful employment outcomes (DeStefano, 1997). The demonstration projects indicated that a consumer-driven model of supported employment was most effective in facilitating successful employment experiences for individuals with severe disabilities. Consumer-driven practices involve the employee and a network of individuals (e.g. service providers, employers, and parents) in the identification and development of appropriate "natural supports" (i.e. human and/or technical accommodations that help employees perform their duties successfully and in a manner that facilitates their integration into the worksite) (Bullock, 1994). These practices differ from traditional models of supported employment that guide the provision of services in many social service agencies. Here, agencies place the responsibility for planning and delivering services on job developers and rely on job coaches to help individuals with disabilities perform their duties successfully through the provision of long-term support (DeStefano, 1997; Hanson, Trach & Shelden, 1997). Thus the transition from a traditional model of supported employment to a consumer-driven approach required major systemic change.

Program Developers at the University of Illinois sought to initiate system-wide change by designing an inservice training program that combined instruction with supervised practical experiences. Two broad goals influenced the content and delivery of the training: to facilitate an understanding of the core program objectives and to encourage the implementation of these objectives in a practical agency context. Eighteen professionals from nine communities throughout the United States were selected to participate in the program as regional teams. The professionals were introduced to a consumer-driven model of supported employment during three training sessions over the course of a year. They were provided with opportunities to implement the model in their agencies through two assignments: a demonstration project and workshop. The former required regional teams to work cooperatively to identify and develop natural supports in the worksite for two individuals with severe disabilities. The latter required teams to disseminate information about natural supports among colleagues within their agencies.

The training program was funded by a grant from the Rehabilitation Services Administration (RSA). In order to comply with RSA stipulations for the provision of funds, an evaluative component was built into the program design. The evaluation examined the relationship between the goals, implementation, and impact of the training. Nine questions served as performance criteria. Six Likert-scale surveys were used to measure performance on seven of the nine criteria, and the design specified that observations should be conducted throughout the training to address the remaining criteria. An examination of the evaluative findings from 1994–1996 revealed that evaluators relied exclusively on the predetermined criteria and methods to assess program functioning.

Evaluative Design

Efforts to understand the quality of the inservice training program required a significant departure from the built-in evaluative component: a flexible design and additional forms of data collection were needed to implement a dialectic process of knowledge construction. Fortunately, RSA stipulations allowed considerable latitude for expansion. Moreover, the program coordinator looked forward to the possibility of gaining greater insight into the nine criteria and other aspects of program functioning. Thus the following plans were made to integrate the built-in evaluative component into a broader effort to pursue an understanding of program quality.

Constructs and criteria. Three constructs were used to guide the pursuit of quality: (a) To what extent is the training effective in introducing professional participants to a consumer-driven model of supported employment? (b) How do professional participants apply a consumer-driven model of supported employment when conducting their demonstration projects and workshops? and (c) What impact does the training have on the professional participants' supported employment practices? The nine criteria remained intact and were embedded into the constructs. However, since predetermined measures would provide only a limited insight into program composition, the criteria were viewed as points of departure for the pursuit of quality. As understanding evolved the existing criteria would be refined and new criteria developed.

Methodology. A range of qualitative methods were combined with survey and observational data to investigate stakeholders' particular program experiences from several frames of references. Interviews were conducted with the four program staff after each training session and two teams of professionals were selected to participate in a case study. The case teams were observed throughout the training and interviewed after each session; moreover, site visits were made to the teams' agencies to assess their progress with the demonstration project.

Implementing the Pursuit

Because it is beyond the scope of this paper to provide a comprehensive account of efforts to understand the quality of the inservice training program, the following discussion focuses upon just one component of the training: the "partner planning times." During the second training session, regional teams were given three 90-minute planning times to discuss their demonstration project and workshop. Teams were asked to evaluate the progress made with each assignment between the first and second sessions, to address any problems or concerns, and to apply the information introduced during the second session to their existing plans. The planning times also gave program staff an opportunity to provide each team with direct support.

The Built-in Evaluative Component

The built-in evaluative component contained one question that related to planning times: "Was the information presented during the training informative and useful?" At the conclusion of the second session the professionals were asked to rate how informative and useful they found each planning time, in terms of the issues discussed and plans made, on a 4-point scale (1 = low, 2 = moderately low, 3 = moderately high, 4 = high). Survey results indicated moderately high mean information and utility ratings across all three planning times. Average information ratings ranged from 3.00 to 3.46, and average utility ratings ranged from 3.23 to 3.38. An examination of individual responses revealed that the majority of professional participants considered the planning times moderately or highly informative and useful. Three professionals gave the first planning time moderately low information ratings and two gave moderately low utility ratings. Only one professional gave low information and utility ratings across the three planning times.

Observational data indicated that the regional teams used the planning time appropriately (i.e. to address problems and develop plans). Yet some teams did not remain on-task throughout the sessions: there were times when little discussion took place and the professionals engaged in various off-task behaviors (e.g. lying on couches, playing the game machines in the lobby, etc.). Observational data was limited, however, as teams were dispersed throughout the conference center, and it was difficult to monitor their behavior.

The Dialectic Model

The dialectic approach was implemented to provide greater insight into the informativeness and utility of the planning times and to further an understanding of the first evaluative construct: To what extent is the training effective in introducing participants to a consumer-driven model of supported employment? Interviews were conducted with the two case teams and four program staff to explore their impressions of the planning times.

Case teams. The teams were asked to evaluate the usefulness of the planning times and to describe the issues discussed and plans made.

Utility. The two professionals in Team 1 agreed that the planning times were the most useful feature of the training as their schedules afforded them little opportunity to work on the assignments between sessions. Indeed, prior to the second session they had not discussed the workshop and were just beginning the demonstration project (i.e. they had identified two individuals to participate in the project). Team 1 therefore needed additional planning time to lay the groundwork for the two assignments. In contrast, Team 2 stated that the planning times were the least useful feature of the training as they met regularly between sessions to work on the assignments and had made considerable progress. Indeed, they arrived at the second session having organized various logistical details for the workshop (i.e. date, venue, cost, marketing strategies etc.). They had also identified two individuals to participate in the demonstration project and were beginning to develop natural supports in the worksite. Team 2 therefore considered additional planning time unnecessary.

Issues and plans. Team 1 used the planning times to discuss the logistics of the workshop (i.e. venue, cost, refreshments, and marketing strategies). However, both professionals pointed out that they had never organized or conducted a presentation before and, consequently, found it difficult to make informed decisions. Determining an appropriate admittance fee for the workshop proved particularly problematic. If the fee was too high they felt it would be prohibitive, but if the fee was too low they felt the training would not be considered worthwhile. These concerns were exacerbated because the professionals worked in a rural community and believed it would be difficult to recruit a large audience. Team 2 used the planning times to determine the content of the workshop and to schedule the day's activities. One professional stated that she organized workshops on a regular basis and was familiar with the plans and procedures necessary to conduct the presentation effectively. This professional also assumed responsibility for implementing all logistical arrangements (i.e. marketing the workshop, arranging the refreshments, printing the handouts, etc.).

Program staff. The staff were asked to discuss the regional teams' use of the planning times, to describe their interactions with team members, and to assess the professionals' understanding of the objectives introduced during the training.

Utility. Program staff indicated that the utility of the planning times was largely contingent upon the teams' experiences as they worked on the assignments between the first and second training sessions. The staff noted that the planning times were particularly beneficial for teams who met infrequently and/or encountered difficulties while implementing the assignments as team members needed an opportunity to develop plans and address areas of concern. In contrast, teams who conducted regular meetings and implemented the assignments successfully had greater difficulty using the planning times productively.

When asked to discuss the off-task behavior observed during the planning times, the staff stated that interpersonal conflicts had arisen within several teams between training sessions. Some team members were unable to develop mutually acceptable plans for the demonstration project and workshop, and others were dissatisfied with the delineation of responsibility as they planned and implemented the assignments. The staff suggested that these difficulties prevented teams from collaborating effectively during the planning times.

Interactions. As the staff described their interactions with the regional teams, it became evident that the professionals did not approach them with questions or concerns as much as they would have liked. The staff were unsure of the factors contributing to this state of affairs, but they wondered if the professionals valued their expertise or were fully invested in the training. Follow-up interviews were conducted with the case teams to investigate this issue. Team 1 stated that they did not interact with the staff during the planning times as they met in their hotel room and were not aware staff support was available. Team 2 approached the staff once, but they considered further support unnecessary since both assignments were well underway. Moreover, like Team 1, they worked in an isolated location and rarely saw the staff.

Understanding. Discussions with the program staff suggested that although the professionals were becoming familiar with consumer-driven practices, many had not grasped "the big picture." The staff mentioned two core training objectives that were proving particularly problematic: (a) the professionals discussed reducing the amount of job coach support or using the job coach as a natural support, rather than developing natural supports to replace the job coach; and (b) the professionals focused exclusively on the worksite when assessing the employee's support needs, rather than looking beyond the worksite to consider the whole environment. The staff identified several factors that may have influenced the professionals' understanding of these objectives. First, some professionals felt their existing practices were sufficiently effective, and others were not convinced natural supports could accommodate the needs of their most challenging clients. Second, professionals working within a traditional model of supported employment did not believe they could effect systemic change within their agencies. Third, teams who made little progress with their demonstration projects were unable to develop understanding through the practical application of the consumer-driven model.

Assessing the Pursuit

Efforts to integrate the prescribed criteria and methods of the built-in evaluative component into a dialectic model of evaluation problematized various design decisions and method choices when seeking to understand the quality of the inservice training program. Survey and observational data alone did not provide sufficient opportunity to explore the experiences of diverse stakeholders, and an exclusive reliance on predetermined criteria to inform the evaluation restricted the extent to which meaning could evolve. Hence, the built-in evaluative component allowed little insight into the complex contingencies that shaped stakeholders' program experiences and, thereby, increased the potential for misunderstanding program conditions. The flexible design and range of qualitative methods associated with the dialectic approach allowed a continuous process of knowledge construction that brought stakeholders' particular program experiences into contact with larger systems of social, political, and institutional relations. This evolving evaluative process contributed to a generic understanding of the inservice training program in a number of ways:

First, greater insight was gained into the evaluative criteria. Interviews with the case teams and program staff extended survey and observational data to provide richer understandings of the different contingencies that influenced the professionals' perceptions of, and experiences during, the planning times.

Second, additional issues and themes associated with the evaluative criteria were identified and investigated. For instance, follow-up interviews were conducted with the case teams to respond to the staffs' account of their interactions with the professionals during the planning times. This unexpected course of investigation provided new insight into the case teams' use of the planning times and prompted a consideration of the differential effectiveness of the planning times for regional teams.

Third, understanding was extended beyond the predetermined criteria as issues and themes emerged that had a broader impact across the three evaluative constructs. The collaborative difficulties program staff observed within several regional teams evolved into a particularly important theme. An additional evaluation question was developed to explore the relationships between team members over the course of the training and to consider the impact of these relationships on the professionals' understanding and implementation of a consumer-driven model of supported employment.

Fourth, perceptions of program composition became more complex throughout the dialectic process. Program staff contributed to the evolution of meaning by suggesting several contingencies that influenced the professionals' understanding of program objectives (i.e. attitudes, agency practices, and implementation experiences). In doing so, the staff not only grounded the program contextually within a wider social and institutional system but also revealed the reciprocal relationship between evaluative constructs (i.e. understanding is necessary for correct implementation, yet implementation enhances understanding). Further investigation was required to explore the impact of these, and other, contingencies on the professionals' program experiences and to examine the interrelationships between evaluative constructs.

UNDERSTANDING QUALITY

The postmodern perspective draws the evaluator's attention to the complex array of contingencies – both locally and in larger systems of social, material, and symbolic relations – that shape program conditions. The evaluation of the inservice training program illustrates the way a dialectic model of evaluation can be employed to explore these contingencies in order to provide a generic sense of program quality. Program composition gained meaning and dimension throughout the dialectic process as the perceptions and experiences of professional participants and program staff were explored. This process provided a view of program composition that was comprised of forces in the larger sociopolitical context, of the diverse institutional and social realities of stakeholders' everyday lives, and of the variety of local program conditions stakeholders encountered during the training. The resulting knowledge base informed evaluative judgments of the multiple meanings of program quality. These judgments required a consideration of the extent to which the training accommodated both the diverse supported employment experiences the professionals brought to the program and the different local program conditions they encountered during the training, in order to maximize their understanding of training objectives and enhance their ability to initiate progressive change within their agencies.

Yet although postmodernism enables the evaluator to construct new and more complex understandings of program conditions, it engenders several challenges when applied to evaluation. The array of interrelated and intractable contingencies that constitute experience prevent the formation of unequivocal knowledge claims, and the ever-changing nature of experience ensures that representations are inevitably open-ended. Moreover, it is difficult to authenticate accounts of everyday life that challenge deeply entrenched notions of reality and uncover the ideological assumptions underlying interpretations of experience which stakeholders themselves may not recognize. This ambiguity and irresolution may cast doubt on both the power of evaluative knowledge claims to challenge conventional wisdom and the ability of evaluative judgments to indicate action that will lead to improved social conditions. Thus as evaluators with postmodern tendencies pursue an understanding of the complexities and contradictions of program quality, which the postmodern perspective reveals, they must explore ways to contend with the radical uncertainty and prohibitive relativism the pursuit may bring.

REFERENCES

Beck, C. (1993). *Postmodernism, pedagogy, and philosophy of education.* [On-line]. Available:http://www.focusing.org/postmod.htm

Bullock, D. D. (1994). *An evaluation of the 1992–1993 cycle of the Systematic Plan for Achieving natural Supports project* (SPANS). Unpublished Manuscript.

Burbules, N. C. (1995). Postmodern doubt and philosophy of education. [On-line]. Available:http://www.focusing.org/postmod.htm

Denzin, N., & Lincoln, Y. S. (1998). Entering the field of qualitative research. In: N. K. Denzin & Y. S. Lincoln (Eds), *Handbook of Qualitative Research* (pp. 1–34). Thousand Oaks, CA: Sage.

DeStefano, L. (1997). Forward. In: J. S. Trach & D. L. Shelden (Eds), *Systematic Plan For Achieving Natural Supports: Evaluation of an Inservice Training Program* (pp. iii–vi). Urbana-Champaign, IL: University of Illinois.

Greene, J. (1997). Evaluation as advocacy. *Evaluation Practice*, *18*, 25–35.

Guba, E., & Lincoln, Y. S. (1998). Competing paradigms in qualitative research. In: N. K. Denzin & Y. S. Lincoln (Eds), *The Landscape of Qualitative Research* (pp. 195–220). Thousand Oaks, CA: Sage.

Hanson, M, R., Trach, J. S., & Shelden, D. L. (1997). An evaluation of the Systematic Plan for Achieving Natural Supports (SPANS): A synthesis of results from cycles I-III. In: J. S. Trach & D. L. Shelden (Eds), *Systematic Plan For Achieving Natural Supports: Evaluation of an Inservice Training Program* (pp. 1–92). Urbana-Champaign, IL: University of Illinois.

Kincheloe, J., & McLaren, P. L. (1998). Rethinking critical theory and qualitative research. In: N. K. Denzin & Y. S. Lincoln (Eds), *The Landscape of Qualitative Research* (pp. 260–299). Thousand Oaks, CA: Sage.

Schwandt, T. A. (1998). Constructivist, interpretivist approaches to human inquiry. In: N. K. Denzin & Y. S. Lincoln (Eds), *The Landscape of Qualitative Research* (pp. 221–259). Thousand Oaks, CA: Sage.

Scriven, M. (1993). *Hard-won lessons in program evaluation.* San Francisco: Jossey-Bass Publishers.

Scriven, M. (1994). The final synthesis. *Evaluation Practice*, *15*(3), 367–382.

Shadish, W. R., Cook, T. D., & Leviton, L. C. (1995). *Foundations of program evaluation.* Thousand Oaks, CA: Sage.

Stake, R. (1998). *The fleeting discernment of quality.* Manuscript submitted for publication.

Stake, R., & Eisner, E. (1998). *Opening remarks.* University of Illinois, Department of Educational Psychology. Urbana-Champaign.

Stake, R., Migotsky, C., Davis, R., Cisneros, E., Depaul, G., Dunbar, C., Farmer, R., Feltovich, J., Johnson, E., Williams, B., & Zurita, M. (1997). The evolving synthesis of program value. *Evaluation Practice*, *18*(2), 89–103.

SECTION II

6. THE RELATIONAL AND DIALOGIC DIMENSIONS OF PROGRAM QUALITY

Jennifer C. Greene

INTRODUCTION

The essential character of program evaluation is the rendering of a judgment of program quality, based on identified program criteria or standards. Evaluators do not just claim to know something about the program evaluated, they also claim to know how good it is. In making these judgmental knowledge claims, evaluators must address several critical issues, issues that directly engage the socio-political dimensions of our craft: Whose program standards will be used in making program quality judgments? How will judgments of program quality then be represented? And for what and whose purposes?

These critical issues are explored in this chapter through the particular lens of participatory evaluation (Greene, 1997a; Whitmore, 1998). As an action and practice-oriented, ideologically self-conscious approach to social inquiry, participatory evaluation offers important perspectives on the value-based meanings and practical implications of program quality judgments. Notably, participatory evaluation focuses attention on the democratizing potential of evaluation to promote pluralism and equity, and on the pivotal importance of evaluative practice that is inclusive and critically respectful (Ryan et al., 1998). How these democratic aims and inclusive principles of practice enlighten judgments of program quality constitutes the discourse of this chapter. A brief statement about

Visions of Quality, Volume 7, pages 57–71.

ISBN: 0-7623-0771-4

participatory evaluation sets the conceptual stage for an instructive case example. Reflections on the example offer both insights and challenges to evaluative renderings of program quality judgments.

ENVISIONING PARTICIPATORY EVALUATION

The case example to be shared was guided by a particular conceptualization of participatory evaluation, which is outlined below (see also Greene, 1997a).

Participatory evaluation intentionally involves diverse program stakeholders in a collaborative, dialogic inquiry process that enables the construction of contextually meaningful knowledge, that engenders the personal and structural capacity to act on that knowledge, and that seeks action that contributes to democratizing social change.[1] Participatory evaluation endeavors to democratize public conversations about important public issues (House & Howe, 1998). Its essential value commitment is to democratic principles of equality, fairness, and justice, to a vision of democratic pluralism (Greene, 1997b).

Defining characteristics of participatory evaluation practice include the following: (a) participating stakeholders assume meaningful decision-making authority, (b) stakeholder participants represent a diversity of stances and interests, and (c) the dialogic inquiry process is at least as important as the inquiry results (Whitmore, 1991). More specifically, in a participatory evaluation, stakeholders share responsibility and authority with each other and with the evaluator for deliberations and decisions about the substantive directions and methodological procedures of the evaluation. What program issues or questions to address, which data gathering and analysis methods to use, how the results are to be interpreted, and what action or change implications to draw are all matters for inclusive dialogue in participatory evaluation. Second, participants in this dialogue intentionally span the full diversity of stakeholder interests, including remote policy makers and funders, on-site managers and staff, intended program beneficiaries and their families, and other community interests (Guba & Lincoln, 1989). Thus, the knowledge claims generated in participatory evaluation importantly incorporate a diversity of perspectives and interests, and thereby, participatory evaluation endeavors to fulfill its commitment to pluralism. And third, there is an emphasis in participatory evaluation on the process of conducting the inquiry, especially on the "social relations of inquiry" (Robinson, 1993) and on these relationships as partly constitutive of the evaluative knowledge constructed (Schwandt, 1997a). In participatory evaluation, that is, knowledge is not extracted from an objective social reality by the evaluator as scientific expert through his/her use of specialized procedures. Rather, knowledge in participatory evaluation is constructed – dialogically and deliberatively (House & Howe, 1998) – as the evaluator engages with stakeholders in respectful, ethically informed conversation about critical aspects of the evaluand (Schwandt, 1997a). In key respects, this dialogue is the evaluation, or at least a significant component thereof.

The case example and subsequent reflections that follow have been crafted to emphasize these defining characteristics of participatory evaluation.

A CASE EXAMPLE OF PARTICIPATORY EVALUATION

Project Context and Initiative

Racism in the employment arena continues to have devastating effects on racial and ethnic minorities in the United States. To be excluded from opportunities for meaningful employment and advancement is to be denied access to the fundamental pathways of esteem and credibility in our society. This participatory inquiry project directly challenged racist employment practices in one local community. Framed as a participatory context evaluation (Stufflebeam, 1983), the project was designed to document the nature and extent of local discriminatory practices in employment for both youth and adults, and to initiate educational activities and other actions that would promote greater understanding of and, longer-term, concrete changes in employment patterns and practices in the community.

The local community in which this project was conducted is not immune from the racism that permeates the cultural mores and norms of the larger society. A state Department of Labor report for 1990 indicated that the county unemployment rate was about 4.3%, a figure substantially less than the overall state average of between 7% and 8% for 1990. However, when disaggregated by racial group, the local employment statistics revealed considerable discrepancies. The 1990 local unemployment rate for Caucasians was the same as the county average or 4.3%. For African Americans the rate was 8.7%, for Native Americans 7.9%, for Latinos/Hispanics 5.0%, and for Asians/Pacific Islanders 2.2%. That is, in this community, about one out of every 11 African American and one out of every 12 Native American adults did not bring home a paycheck in 1990.

There had also been specific instances of racism in employment in recent years. The following vignette captures one of these much-publicized instances involving an African American teen seeking a job:

"No, I'm sorry, we're no longer accepting applications", said the fast food restaurant manager to the teenager inquiring about a part-time counter job.

"But, your ad said that applications would be accepted through the end of this week", protested the teen.

"Well, I think we already have enough. In fact, we're kinda swamped with applications. Maybe next time", said the manager as she turned and walked back to her office.

Following this incident in late 1989, several local community-based organizations, in tandem with local affiliates of state organizations like the state Department of Labor and Job Services, came together to form the Coalition Against Racism in Employment. Most of these organizations had long been working to challenge racial discrimination in the workplace. By joining forces they hoped to have a stronger voice and greater effectiveness. From the beginning, the Coalition conducted a variety of educational and advocacy activities. Unintentionally, these activities had fostered a climate of hostility and adversarial relationships between the Coalition and members of the local business community.

It was in this context that the present participatory inquiry project was envisioned and implemented during 1993–94. One explicit project aim was to shift the adversarial relationship between the Coalition and local business groups to a cooperative partnership. It was hoped that joint participation on the project team would facilitate such a shift. Key initiative for the project came from the Coalition. I worked with Coalition members to write the project proposal, because funding for the project needed to pass through a university, in part to promote undergraduate student involvement in community-based endeavors. But, as envisioned, the project was to have a community, not a university, home.

Project Description

Leadership

The project was co-directed by myself, a university-based Caucasian woman, and a member of the Coalition, an African American woman with considerable longevity within the community. Several other members of the Coalition joined the two co-directors in an unofficial steering committee to help get the project going.

Participatory Inquiry Team

The first important task of the project was to constitute the participatory inquiry team. Approximately eight meetings were convened by the project co-directors between late August and early October of 1993 primarily for this purpose. Most of these meetings were held with the Coalition or with the informal project steering committee. As the community-based initiator of this project, the Coalition had considerable authority over critical project decisions and directions, like inquiry team membership. For example, while the Coalition agreed with the categories of team membership offered in the project proposal, they wanted – and attained – a reallocation of membership slots so that more team members came from the community than from the university.

The team membership categories and slots agreed to by the Coalition were as follows: the Coalition, which itself was comprised of various community-based organizations, 3–4 representatives; teens and adults of color who were seeking or holding jobs, at least 2 representatives; area employers, 2–3 representatives; university undergraduate students recruited from a fall course offering on racism, 5 representatives; university faculty, 1–3 representatives (myself as co-director and possibly one or both of the faculty teaching the racism course)[2] and 1 graduate student project assistant. Major criteria set forth by the Coalition for identifying and recruiting these representatives were relevant experience, awareness that employment discrimination exists, and a commitment to help effect change. More specific criteria for each membership category were also established.

In addition to defining team membership, these initial planning meetings addressed the challenges of recruitment. Recruitment of four volunteers from the Coalition and five interested undergraduate students was accomplished efficiently and effectively. Identifying community-based teens and adults of color and identifying interested area employers was considerably more difficult. In fact, the challenges of recruitment for (and later retention in) team membership remained difficult throughout the project, effecting significant delays in project start-up and subsequent implementation. The philosophy of participatory evaluation insists that multiple, diverse perspectives be heard and incorporated into significant inquiry decisions and directions. Proceeding without some of those perspectives is therefore counter to the essential core of participation. In our project, it took a long time to secure all of the perspectives deemed critical to our work, and in this process we used up precious months of what was only one year together.

At the time of the first project team meeting in early October 1993, one teen and one adult of color but no employers had been recruited, in addition to the Coalition members and undergraduate students. By the next meeting, one additional adult and one additional youth of color, but still no employers, had been recruited. In fact, it wasn't until the December 8th meeting that one area employer joined the team. And it was at this point that the team stopped further recruitment efforts, having allocated considerable time and energy throughout the fall to this effort.

Finally, recruitment is one challenge, active participation and retention another. Even though this participatory inquiry team was eventually successfully constituted, actual participation was uneven and inconsistent. Throughout the winter and spring periods when the project was most heavily engaged in its data collection and analysis tasks, only some team members evinced meaningful and sustained participation (by regularly attending meetings and by conducting work outside team meetings). These included the project co-directors, some but not all of the Coalition members, some but not all of the undergraduates, only one adult from the teen/adult group of color, our lone employer representative, and three graduate students from the university who had volunteered for the project due to personal interest and commitment. Among the difficulties invoked by uneven participation in this particular case was the frequent imbalance of team members present at any given meeting. University-based team members were frequently in the majority, despite clear intentions to ground the project in a community home and to site project ownership unequivocally in the community.

Major Activities

So, what did we do in this project? We had 22 team meetings between October 1993 and July 1994, and we pursued a small follow-up effort over the summer and fall. We conducted a pilot version of the inquiry activities really needed to fulfill the project's evaluation objectives. We planted some seeds of activism that might sprout in the future. And we struggled to understand and work coilaboratively and respectfully with each other. Some elaboration of these activities is offered in this section. The project's ambitious evaluation questions were the following: What is the nature and extent of discriminatory practices in employment – including job access, retention, and mobility – in this local community? What are the contextual, structural, and institutionalized influences on employment discrimination in the community, for example, the ways in which tracking in the school system influences racism in employment practices?

The team initiated work on these evaluation questions by first each generating our own concerns and questions about racism in employment, in effect, grounding these general questions into specific, contextualized concerns. Examples of these follow: What is it like for a young person to be repeatedly turned down in seeking a job – what does that do to her/his sense of self? How do the employment/business sectors think about the potential pool of employees? What are employers' meanings and understandings of diversity? If employers are hiring diverse people, what are they doing to change the workplace culture to encourage retention of these same people? Are instances of discrimination different in different job sectors? These were discussed, as were methodological ideas about how to gather information related to the question and about what would constitute appropriate inquiry samples. In addition to issues of team membership and recruitment, these discussions were the primary agenda of the group during the first few team meetings.

In late October, I aggregated and organized these discussions into a preliminary chart that had three columns: (1) "*What* do we want to learn about or better understand?" which included the specific questions generated by team members; (2) "From *whom*?" which addressed sampling ideas; and (3) "*How*? With what methods?" which included team members' methods suggestions. This chart worked effectively to anchor but not constrain team interactions about the substantive agenda of our work together. I updated the chart twice over the ensuing month, to reflect team discussions and decisions. For the December 8th meeting, I also prepared a draft set of "instruments and sampling plans" that both reflected and elaborated upon team work during the fall. During the two December meetings, these drafts were discussed and revised.

The December holiday break was the occasion for several key project events. First, during the break, several team members (myself included) took a step back from the project and gasped at the scope of our ambitions, especially compared to our meager resources. In our first January meeting, we who had gasped encouraged the group to reframe our work as a pilot effort. With discussion, the team agreed, deciding to retain all study components but only conduct a small pilot evaluative inquiry for each. Second, the tensions between the Coalition's keen desire to own the project and the dominant presence of the university in most team meetings became more apparent. These tensions were to accompany the project for the remainder of its existence, working to significantly dissipate energies and commitments. And third, a serious racial incident in a downtown retail store during the peak holiday shopping period infiltrated the team's dynamics when we resumed meeting in mid-January. In this incident, an African American family had been treated with suspicion and a lack of courtesy by the white shop owner when they had entered and tried to browse around his gift shop. This store is located in a retail area which has a local history of not welcoming people of color either as employees or as consumers. Small demonstrations protesting the incident and media attention also occurred.

In the first project meeting after the holidays, different perspectives on and understandings of this incident quickly occluded other agenda items. The primary participants in the impassioned exchange that occurred were a leader of the Coalition, who was an African American woman, and our lone employer representative, who was a white man. Both at the time and since then, I have viewed this exchange as highly consonant with the values and premises of participatory evaluation and as a significant opportunity for the project. The intentional diversity of the participatory inquiry team invited tensions and differences from the world outside into our meeting rooms and interactions. And I also wondered, "How can the project enter into this situation and use the energies already stirred up to facilitate our data gathering and dialogue-action plans?" Related to the latter idea, I participated in the first of several community meetings convened to follow up on the shop incident, but was unable to forge a meaningful connection to our project, partly because these follow-up gatherings suffered many delays, due to weather and politics alike.

Then the Coalition-university tensions split wide open in late February. This happened in conjunction with continued follow-up to the racial holiday shopping incident. Specifically, the Coalition had requested and been granted time to speak with a downtown business association, a group that included the shop owner involved in the incident. The Coalition's intention here was primarily educational, to begin to sensitize business owners about the racial dynamics involved. Preparation of the Coalition's presentation to this group became the controversial subject of discussion at February project meetings. Some non-Coalition project team members were concerned that too close an alliance between the Coalition and this project would jeopardize meaningful data collection from employers and subsequent credibility of project findings and actions. Coalition members were deeply disturbed at this disassociation of the project from the Coalition. Feeling already vulnerable, they felt further unsupported and attacked by this disassociation. I pursued this by attending a March Coalition meeting, with the primary agenda of better understanding the concerns and perspectives of Coalition members. While this helped to smooth personal relationships between Coalition members and myself, within the project the overall rift had split too far. From this point on, many Coalition members of the team were effectively disengaged from the project's activities and commitments.

In mid-May (with the university students leaving for the summer and the July project deadline looming), I decided in consultation with other team members that it was time to bring the evaluative phase of the project to a close and shift to the action phase.[3] Each of the project work groups prepared a draft summary of their pilot work and findings. Then, the final two team meetings in June involved a discussion of how to effect the shift to action. These discussions addressed the scope and defensibility of our pilot evaluation results, how to synthesize and report them, and what action initiatives to undertake.

With the formal ending of the project in July, these action alternatives did not fully materialize. Nonetheless, there were two important, action-oriented follow-up activities conducted over the summer and fall. First, a small group from the project team voluntarily pursued: (a) preparation and dissemination of a final report from the project, and (b) preparation and dissemination of shorter news articles about our work in the local press, including a short piece in a local business newsletter. Second, members of the Coalition pursued the making of a video on racism in the workplace with area youth, including some of the youth who had participated in this project (as team members and as interview respondents). Final closure of the project was effected with two additional follow-up meetings in the fall of 1994.

An analysis of why this participatory context evaluation of racism in employment was not as successful as intended is highly instructive for evaluators interested in the practice of participatory evaluation (see Greene, in press). More relevant for the present discussion are the connections between the key relational and dialogic dimensions of participatory evaluation and evaluative judgments about program quality, both in participatory evaluation and in evaluation more broadly, as addressed in the reflections that follow.

THE RELATIONAL AND DIALOGIC DIMENSIONS OF EVALUATIVE CLAIMS ABOUT QUALITY

So, what are the connections between this case example of participatory evaluation and the fundamental evaluative enterprise of judging program quality? How does this modest example inform this lofty endeavor?

In my view, what is most instructive about this example is that it focuses attention on particular dimensions and representational forms of program quality that are very important to contemporary discourse about public programs and policies in the United States. These dimensions and forms comprise the democratic character of the relationships and dialogue that distinguish interactions among program stakeholders in a program context. That is, this example underscores the centrality of stakeholder relationships and dialogue to evaluative claims about program quality. While central to participatory evaluation (and to this particular evaluative context), these relational and dialogic dimensions of program quality are less salient in or even absent from other ways of evaluating program quality. The argument being advanced here is that these other evaluative judgments are thereby importantly incomplete and insufficient.

QUALITY JUDGMENTS IN PARTICIPATORY EVALUATION

Participatory evaluation is distinguished by the intentional inclusion of multiple, diverse stakeholder interests in the evaluative process. This commitment to inclusion is grounded in the political pluralism that characterizes most contemporary evaluation contexts, certainly evaluations of public programs that address persistent social problems (House & Howe, 1998; Ryan et al., 1998). In such contexts, multiple stakeholders have legitimate claims to a voice, and participatory evaluation is committed to including all legitimate voices, and to forging respectful and equitable relationships among them, at least within the inquiry process. Participatory evaluation, that is, attends carefully to the "social relations of inquiry" (Robinson, 1993), which collectively configure parallel relations of authority and power. In theory, a participatory inquiry team learns how to confront and address, if not redress, power inequities within the group, to dialogue across differences within the relatively safe boundaries of team membership, and then to transfer this learning to outside roles and responsibilities. Stakeholder inclusiveness, equity of voice through meaningful participation, and reciprocal learning and respect through dialogue are the intended social relations of participatory evaluation.

In participatory evaluation therefore, multiple stakeholder interests must be included and respectful democratizing dialogue among those interests must be pursued in order to render meaningful evaluative claims about the evaluand. Evaluative claims that exclude some interests or that have been developed and presented through a monologic versus dialogic process (Abma, 1998; Coulter, 1999) are neither complete nor warranted. The inclusion of pluralistic interests and the engagement of stakeholders in democratic deliberation about these interests are thus constitutive of warranted evaluative claims in participatory evaluation. Moreover, the substance or content of evaluative claims in participatory evaluation encompasses issues of inclusion and pluralism, and of dialogue and deliberation. That is, because the process of participatory evaluation includes multiple stakeholder interests and invites meaningful and respectful conversations among them, the substance of participatory evaluative claims necessarily addresses the presence and the quality of the same dimensions in the evaluand.

Quality Judgments in this Participatory Evaluation Case Example

In the present case, the project inquiry team was indeed intentionally constituted to include diverse perspectives and interests, considerable effort was expended to forge equitable and respectful relationships among team members, and the designed mode of stakeholder interaction was dialogue. This was most evident in the intensive stakeholder recruitment conducted, in the careful planning of project roles and responsibilities, and in the repeated efforts to insure community ownership for the project. In the present case, the diversity of participating stakeholders generated numerous instances of tension and debate within the team. Again in theory, such instances are expected and even welcomed as micro-level manifestations of the macro-level issues being addressed in the project, in our case, racism. In theory, respectful stakeholder dialogue enables reciprocal learning and the development of shared understandings. In the practice of our particular case, however, these instances of tension and debate blocked the team's constructive progress at several critical points. Dialogue gave way to rancorous debate. Respect gave way to rejection. The social relations of our inquiry fractured and broke down. And thereby, our subsequent evaluative claims – of both knowledge and judgment – became diminished and less warranted.

Also related to the social relations of inquiry – notably power and authority – in this project were the ongoing tensions over project ownership. As recounted in the project description above, these tensions surfaced and persisted despite the marked lack of debate over project ownership and authority. From the outset, the project was cast as a community-based endeavor, inspired by the activism of the Coalition. The university's role was to provide support and resources. And no one disputed this allocation of responsibility, authority, or ownership. Tensions surfaced nonetheless for the multiple reasons described above, especially the frequent dominance of university-based team members at project meetings and the split between the project and the Coalition during the aftermath of the holiday shopping racial incident – all within a historical context of ongoing, at times acerbic "town-gown" disputes. While this participatory evaluation intentionally intended to broaden and democratize ownership and decision authority over the project, the actual pattern of participation and power served primarily to reinforce existing power inequities. Discourse gave way to disengagement, engagement to distrust. And thereby, our subsequent evaluative claims – of both knowledge and judgment – became diminished and less warranted.

Ultimately, in our participatory context evaluation of racism in local employment practices, the loss of some perspectives and interests from the evaluative process meant the loss of those perspectives and interests from the ongoing evaluation conversation, from the evaluative results that were reached, and from the evaluation's judgments about quality. Because the African-American community-based activist members of the Coalition disengaged from the evaluation process, their perspectives and interests were not meaningfully or effectively included in the ensuing dialogue or the subsequent evaluative claims to know and judge the extent and nature of racism in local employment. In this particular case, this loss was severe if not catastrophic, as the very focus of the evaluation was on a socio-political issue of equity, fairness, and inclusion.

Quality Judgments in Evaluation More Broadly

Participatory evaluation underscores the importance of attending to the democratic character of stakeholder relationships and dialogue in making judgments about program quality; but not just in contexts explicitly concerned with issues of equity and fairness, rather in all evaluation contexts. Human endeavors of all kinds fundamentally engage these issues, and so should our evaluative lenses.

The work of Tineke Abma in the Netherlands well illustrates such an evaluative lens. For example, her evaluation of a Palliative Care project for terminal cancer patients used a "responsive approach [, which] takes plurality as a point of departure to facilitate a dialogue between those with an interest in the evaluated program. Criteria [for judging quality] are [derived] from the various, and sometimes conflicting, values and meanings of a great many individuals with an interest in the evaluated program" (1998, p. 435). As one means to explicitly promote dialogue, Abma intentionally crafted a dialogic reporting process "about the meaning of palliative care and the consultative process" (p. 437). In this process, written reports offered multiple stories and perspectives, especially including stories from those "who felt more marginalized and whose voice was not (yet) heard" (p. 442). Reports "did not present one final interpretation about the quality of palliative care – but were composed of diverging experiences and multiple perspectives" (p. 446). Reports also offered "polyvocality", meaning the professional jargon of health professionals ("anti-cancer drugs") alongside the plain talk of patients ("poison"), highlighted by the use of different literary genres. Reports did not offer conclusions and recommendations but rather a recursive plot that invited engagement and commentary. Moreover, Abma structured several opportunities for various stakeholders to comment and converse with one another about the reports. In these conversations, stakeholders did not reach consensus on quality judgments about the program, but did gain insight and understanding into each other's criteria and standards of success, which over time, stimulated further reflection and conversation.

Abma's (1998) responsive evaluation, in both form and substance, thus centrally addressed the democratic character of people's relationships and interactions with one another in this health program setting. And judgments of program quality were engaged through dialogue and with respect for multiple, diverse perspectives.

Similarly, all of our evaluative endeavors should wonder how well a program advances multiple stakeholder interests through open conversation. All of our evaluative endeavors should not offer closed claims to know but open invitations for further dialogue about our differences, our continuing conflicts, as well as our shared understandings and aspirations. All of our evaluative endeavors should

> Affirm . . . a social world accepting of tension and conflict. What matters is an affirmation of energy and the passion of reflection in a renewed hope of common action, of face-to-face encounters among friends and strangers, striving for meaning, striving to understand. What matters is a quest for new ways of living together, of generating more and more incisive and inclusive dialogues (Greene, 1994, p. 459).

REPRISE

The insights from participatory evaluation thus suggest the following responses to the critical questions posed at the outset of this paper regarding evaluative judgments of program quality:

(1) Whose program standards will be used in making program quality judgments? Judgments of program quality should be based on multiple program standards that inclusively represent the interests of all legitimate stakeholders. Judgments of program quality should address the degree to which, and ways in which, the evaluand promotes or advances diversity and inclusivity. Judgments of program quality are inherently multiplistic in form and content.

(2) How will judgments of program quality then be represented? Evaluative judgments of program quality should be represented dialogically, involving diverse legitimate stakeholders in open, respectful conversation with one another. Program quality is thereby not a static, fixed property of an evaluand, but rather an opportunity for deliberation about various aspects of the evaluand held dearly by diverse stakeholders, an opportunity for "face-to-face encounters among friends and strangers, striving for meaning, striving to understand" (Greene, 1994, p. 459).

(3) And for what and whose purposes? If evaluative judgments of program quality address the evaluand's legitimization of diversity and inclusivity and if such judgments offer opportunities for meaningful conversation among diverse stakeholders, embracing "a renewed hope of common action", then evaluation is being conducted in the interests of democratic

pluralism. In addition, social inquiry itself becomes reframed – from a separate activity, a disjuncture in people's lives, to an integrative, connective activity that respects, even supports the daily endeavors and the dreams that sustain them for people and their human practices (Abma, 1998; Schwandt, 1997b).

NOTES

1. Social change, however, is viewed with an evolutionary rather than a revolutionary framework, one marked by progressions of small individual and collective steps (Weiss & Greene, 1992).

2. In fact, only one of these two faculty members participated in the project, and his participation was uneven.

3. From the outset this project had action objectives in addition to evaluation objectives. The two action objectives were: (a) To use the results of the inquiry to develop concrete action strategies toward elimination of workplace discrimination in the community, via a process of critical dialogue involving representatives of all key constituencies; and (b) to develop and disseminate educational materials that present the results of this inquiry project and that engage target audiences in further dialogue about promising change strategies.

4. The data gathering and analysis activities of the work groups had been unevenly implemented during the winter and spring months.

REFERENCES

Abma, T. A. (1998). Text in an evaluative context: Writing for dialogue. *Evaluation*, *4*, 434–454.

Coulter, D. (1999). The epic and the novel: Dialogism and teacher research. *Educational Researcher*, *28*(3), 4–13.

Greene, J. C. (in press). In the rhythms of their lives. In: M. Levin & Y. S. Lincoln (Eds), *Social Research for Social Change*. Thousand Oaks, CA: Sage.

Greene, J. C. (1997a). Participatory evaluation. In: R. Stake (Series Ed.) & L. Mabry (Vol. Ed.), *Advances in Program Evaluation: Vol. 3. Evaluation and the Post-Modern Dilemma*. (pp. 171–189). Greenwich, CT: JAI Press.

Greene, J. C. (1997b). Evaluation as advocacy. *Evaluation Practice*, *18*, 25–35.

Greene, M. (1994). Epistemology and educational research: The influence of recent approaches to knowledge. In: L. Darling-Hammond (Ed.), *Review of Research in Education*, *20*, 423–464.

Guba, E. G., & Lincoln, Y. S. (1989). *Fourth generation evaluation*. Thousand Oaks, CA: Sage.

House, E. R., & Howe, K. R. (1998). *Deliberative democratic evaluation*. Paper presented at the annual meeting of the American Evaluation Association, Chicago.

Robinson, V. (1993). *Problem based methodology: Research for the improvement of practice*. Oxford: Pergamon.

Ryan, K., Greene, J., Lincoln, Y., Mathison, S., & Mertens, D. (1998). Advantages and challenges of using inclusive evaluation approaches in evaluation practice. *American Journal of Evaluation*, *19*, 101–122.

Schwandt, T. A. (1997a). Evaluation as practical hermeneutics. *Evaluation*, *3*, 69–83.

Schwandt, T. A. (1997b). Reading the "problem of evaluation" in social inquiry. *Qualitative Inquiry, 3*, 4–25.

Stufflebeam, D. L. (1983). The CIPP model for program evaluation. In: G. F. Madaus, M. Scriven, & D. L. Stufflebeam (Eds), *Evaluation Models* (pp. 117–141). Boston: Kluwer Nijhoff.

Weiss, H., & Greene, J. C. (1992). An empowerment partnership for family support and education programs and evaluation. *Family Science Review, 5*, 131–148.

Whitmore, E. (1991). *Evaluation and empowerment: It's the process that counts*. Networking Bulletin: Empowerment and Family Support (Vol. 2, pp. 1–7). Ithaca, New York: Cornell Empowerment Project.

Whitmore, E. (1998) (Ed.). Understanding and practicing participatory evaluation. *New Directions for Evaluation, 80*. San Francisco: Jossey-Bass.

7. EMPOWERMENT EVALUATION: THE PURSUIT OF QUALITY

David M. Fetterman

INTRODUCTION

Empowerment evaluation is an approach to evaluation that places the primary responsibility for the pursuit of quality in the program staff and participants' hands. It is specifically designed to help program staff members and participants learn to recognize quality and represent it in their programs. This discussion presents an overview of empowerment evaluation and in the process highlights the role of the evaluation coach in helping people assess and improve the quality of their programs. In addition, this discussion highlights the fundamental role of the group in conducting a self-evaluation.

BACKGROUND

Empowerment evaluation has been adopted in higher education, government, inner-city public education, non-profit corporations, and foundations throughout the United States and abroad. A wide range of program and policy sectors use empowerment evaluation, including substance abuse prevention, HIV prevention, crime prevention, welfare reform, battered women's shelters, agriculture and rural development, adult probation, adolescent pregnancy prevention, tribal partnership for substance abuse, self-determination and individuals with disabilities, doctoral programs, and accelerated schools. Descriptions of programs that use empowerment evaluation appear in *Empowerment Evaluation: Knowledge*

Visions of Quality, Volume 7, pages 73–106.

ISBN: 0-7623-0771-4

and Tools for Self-assessment and Accountability (Fetterman, Kaftarian & Wandersman, 1996). *Foundations of Empowerment Evaluation* (Fetterman, 2000) provides additional insight into this new evaluation approach, including information about how to conduct workshops to train program staff members and participants to evaluate and improve program practice (see also Fetterman, 1994a, b, and 1996a). In addition, this approach has been institutionalized within the American Evaluation Association[1] and is consistent with the spirit of the standards developed by the Joint Committee on Standards for Educational Evaluation (Fetterman, 1995; Joint Committee on Standards for Educational Evaluation, 1994).[2]

Empowerment evaluation is the use of evaluation concepts, techniques, and findings to foster improvement and self-determination. It employs both qualitative and quantitative methodologies. Although it can be applied to individuals, organizations,[3] communities, and societies or cultures, the focus is usually on programs.

Empowerment evaluation is attentive to empowering processes and outcomes. Zimmerman's work on empowerment theory provides the theoretical framework for empowerment evaluation. According to Zimmerman (in press):

> A distinction between empowering processes and outcomes is critical in order to clearly define empowerment theory. Empowerment processes are ones in which attempts to gain control, obtain needed resources, and critically understand one's social environment are fundamental. The process is empowering if it helps people develop skills so they can become independent problem solvers and decision makers. Empowering processes will vary across levels of analysis. For example, empowering processes for individuals might include organizational or community involvement, empowering processes at the organizational level might include shared leadership and decision making, and empowering processes at the community level might include accessible government, media, and other community resources.
>
> Empowered outcomes refer to operationalization of empowerment so we can study the consequences of citizen attempts to gain greater control in their community or the effects of interventions designed to empower participants. Empowered outcomes also differ across levels of analysis. When we are concerned with individuals, outcomes might include situation specific perceived control, skills, and proactive behaviors. When we are studying organizations, outcomes might include organizational networks, effective resource acquisition, and policy leverage. When we are concerned with community level empowerment, outcomes might include evidence of pluralism, the existence of organizational coalitions, and accessible community resources (pp. 8–9).

Empowerment evaluation has an unambiguous value orientation – it is designed to help people help themselves and improve their programs using a form of self-evaluation and reflection. This approach to evaluation has roots in community psychology and action anthropology. It has also been influenced by action research and action evaluation. Program participants – including clients –

conduct their own evaluations; an outside evaluator often serves as a coach or additional facilitator depending on internal program capabilities.

ROLE AND CONTEXT REDEFINITION: A FOCUS ON THE GROUP AND THE ENVIRONMENT

Zimmerman's (in press) characterization of the community psychologist's role in empowering activities is easily adapted to the empowerment evaluator:

> An empowerment approach to intervention design, implementation, and evaluation redefines the professional's role relationship with the target population. The professional's role becomes one of collaborator and facilitator rather than expert and counselor. As collaborators, professionals learn about the participants through their culture, their world-view, and their life struggles. The professional works *with* participants instead of advocating *for* them. The professional's skills, interest, or plans are not imposed on the community; rather, professionals become a resource for a community. This role relationship suggests that what professionals do will depend on the particular place and people with whom they are working, rather than on the technologies that are predetermined to be applied in all situations. While interpersonal assessment and evaluation skills will be necessary, how, where, and with whom they are applied can not be automatically assumed as in the role of a psychotherapist with clients in a clinic (pp. 8–9).

Empowerment evaluation also requires sensitivity and adaptation to the local setting. It is not dependent upon a predetermined set of technologies. Empowerment evaluation is necessarily a collaborative group activity, not an individual pursuit. An evaluator does not and cannot empower anyone; people empower themselves, often with assistance and coaching. This process is fundamentally democratic in the sense that it invites (if not demands) participation, examining issues of concern to the entire community in an open forum.

As a result, the context changes: the assessment of a program's value and worth is not the endpoint of the evaluation – as it often is in traditional evaluation – but is part of an ongoing process of program improvement. This new context acknowledges a simple but often overlooked truth: that merit and worth are not static values. Populations shift, goals shift, knowledge about program practices and their value change, and external forces are highly unstable. By internalizing and institutionalizing self-evaluation processes and practices, a dynamic and responsive approach to evaluation can be developed to accommodate these shifts. Both value assessments and corresponding plans for program improvement – developed by the group with the assistance of a trained evaluator – are subject to a cyclical process of reflection and self-evaluation. Program participants learn to continually assess their progress toward self-determined goals, and to reshape their plans and strategies according to this assessment. In the process, self-determination is fostered, illumination generated, and liberation actualized.

Value assessments are also highly sensitive to the life cycle of the program or organization. Goals and outcomes are geared toward the appropriate developmental level of implementation. Extraordinary improvements are not expected of a project that will not be fully implemented until the following year. Similarly, seemingly small gains or improvements in programs at an embryonic stage are recognized and appreciated in relation to their stage of development. In a fully operational and mature program, moderate improvements or declining outcomes are viewed more critically.

ROOTS, INFLUENCES, AND COMPARISONS

Empowerment evaluation has many sources. The idea first germinated during preparation of another book, *Speaking the Language of Power: Communication, Collaboration, and Advocacy* (Fetterman, 1993). In developing that collection, I wanted to explore the many ways that evaluators and social scientists could give voice to the people they work with and bring their concerns to policy brokers. I found that, increasingly, socially concerned scholars in myriad fields are making their insights and findings available to decision makers. These scholars and practitioners address a host of significant issues, including conflict resolution, the drop-out problem, environmental health and safety, homelessness, educational reform, AIDS, American Indian concerns, and the education of gifted children. The aim of these scholars and practitioners was to explore successful strategies, share lessons learned, and enhance their ability to communicate with an educated citizenry and powerful policy-making bodies. Collaboration, participation, and empowerment emerged as common threads throughout the work and helped to crystallize the concept of empowerment evaluation.

Empowerment evaluation has roots in community psychology, action anthropology, and action research. Community psychology focuses on people, organizations, and communities working to establish control over their affairs. The literature about citizen participation and community development is extensive. Rappaport's (1987) *Terms of Empowerment/Exemplars of Prevention: Toward a Theory for Community Psychology* is a classic in this area. Sol Tax's (1958) work in action anthropology focuses on how anthropologists can facilitate the goals and objectives of self-determining groups, such as Native American tribes. Empowerment evaluation also derives from collaborative and participatory evaluation (Choudhary & Tandon, 1988; Oja & Smulyan, 1989;

Papineau & Kiely, 1994; Reason, 1988; Shapiro, 1988; Stull & Schensul, 1987; Whitmore, 1990; Whyte, 1990).

Empowerment evaluation has been strongly influenced by and is similar to action research. Stakeholders typically control the study and conduct the work in action research and empowerment evaluation. In addition, practitioners empower themselves in both forms of inquiry and action. Empowerment evaluation and action research are characterized by concrete, timely, targeted, pragmatic orientations toward program improvement. They both require cycles of reflection and action and focus on the simplest data collection methods adequate to the task at hand. However, there are conceptual and stylistic differences between the approaches. For example, empowerment evaluation is explicitly driven by the concept of self-determination. It is also explicitly collaborative in nature. Action research can be either an individual effort documented in a journal or a group effort. Written narratives are used to share findings with colleagues (see Soffer, 1995). A group in a collaborative fashion conducts empowerment evaluation, with a holistic focus on an entire program or agency. Empowerment evaluation is never conducted by a single individual. Action research is often conducted on top of the normal daily responsibilities of a practitioner. Empowerment evaluation is internalized as part of the planning and management of a program. The institutionalization of evaluation, in this manner, makes it more likely to be sustainable rather than sporadic. In spite of these differences, the overwhelming number of similarities between the approaches has enriched empowerment evaluation.

Another major influence was the national educational school reform movement with colleagues such as Henry Levin, whose Accelerated School Project (ASP) emphasizes the empowerment of parents, teachers, and administrators to improve educational settings (Levin, 1996). We worked to help design an appropriate evaluation plan for the Accelerated School Project that contributes to the empowerment of teachers, parents, students, and administrators (Fetterman & Haertel, 1990). The ASP team and I also mapped out detailed strategies for district-wide adoption of the project in an effort to help institutionalize the project in the school system (Stanford University and American Institutes for Research, 1992).

Dennis Mithaug's (1991, 1993) extensive work with individuals with disabilities to explore concepts of self-regulation and self-determination provided additional inspiration. We completed a 2-year Department of Education-funded grant on self-determination and individuals with disabilities. We conducted research designed to help both providers for students with disabilities and the students themselves become more empowered. We learned about self-determined behavior and attitudes and environmentally related features of self-determination by listening to self-determined children with disabilities and their providers. Using specific concepts and behaviors extracted from these case studies, we developed a behavioral checklist to assist providers as they work to recognize and foster self-determination.

Self-determination, defined as the ability to chart one's own course in life, forms the theoretical foundation of empowerment evaluation. It consists of numerous interconnected capabilities, such as the ability to identify and express needs, establish goals or expectations and a plan of action to achieve them, identify resources, make rational choices from various alternative courses of action, take appropriate steps to pursue objectives, evaluate short- and long-term results (including reassessing plans and expectations and taking necessary detours), and persist in the pursuit of those goals. A breakdown at any juncture of this network of capabilities – as well as various environmental factors – can reduce a person's likelihood of being self-determined. (See also Bandura, 1982, concerning the self-efficacy mechanism in human agency.)

A pragmatic influence on empowerment evaluation is the W. K. Kellogg Foundation's emphasis on empowerment in community settings. The foundation has taken a clear position concerning empowerment as a funding strategy:

> We've long been convinced that problems can best be solved at the local level by people who live with them on a daily basis. In other words, individuals and groups of people must be empowered to become change-makers and solve their own problems, through the organizations and institutions they devise. . . . Through our community-based programming, we are helping to empower various individuals, agencies, institutions, and organizations to work together to identify problems and to find quality, cost-effective solutions. In so doing, we find ourselves working more than ever with grantees with whom we have been less involved – smaller, newer organizations and their programs (W. K. Kellogg Foundation, 1992, p. 6).

Its work in the areas of youth, leadership, community-based health services, higher education, food systems, rural development, and families and neighborhoods exemplifies this spirit of putting "power in the hands of creative and committed individuals – power that will enable them to make important changes in the world" (p. 13). For example, one project – Kellogg's Empowering Farm Women to Reduce Hazards to Family Health and Safety on the Farm – involves a participatory evaluation component. The work of Sanders, Barley, and Jenness (1990) on cluster evaluations for the Kellogg Foundation also highlights the value of giving ownership of the evaluation to project directors and staff members of science education projects.

These influences, activities, and experiences form the background for empowerment evaluation. An eloquent literature on empowerment theory by Zimmerman (in press), Zimmerman, Israel, Schulz and Checkoway (1992), Zimmerman and Rappaport (1988), and Dunst, Trivette and LaPointe (1992), as discussed earlier, also informs this approach. A brief discussion about the pursuit of truth and a review of empowerment evaluation's three steps and various facets will illustrate its wide-ranging application.

PURSUIT OF TRUTH AND HONESTY

Empowerment evaluation is guided by many principles. One of the most important is a commitment to truth and honesty – this has significant positive implications concerning the pursuit of quality. This is not a naive concept of one absolute truth, but a sincere intent to understand an event in context and from multiple world-views. The aim is to try and understand what's going on in a situation from the participant's own perspective as accurately and honestly as possible and then proceed to improve it with meaningful goals and strategies and credible documentation. There are many checks and balances in empowerment evaluation, such as having a democratic approach to participation – involving participants at all levels of the organization, relying on external evaluators as critical friends, and so on.

Empowerment evaluation is like a personnel performance self-appraisal. You come to an agreement with your supervisor about your goals, strategies for accomplishing those goals, and credible documentation to determine if you are meeting your goals. The same agreement is made with your clients. If the data are not credible, you lose your credibility immediately. If the data merit it at the end of the year, you can use it to advocate for yourself. Empowerment evaluation applies the same approach to the program and community level. Advocacy, in this context, becomes a natural by-product of the self-evaluation process – if the data merit it. Advocacy is meaningless in the absence of credible data. In addition, external standards and/or requirements can significantly influence any self-evaluation. To operate without consideration of these external forces is to proceed at your own peril. However, the process must be grounded in an authentic understanding and expression of everyday life at the program or community level. A commitment to the ideals of truth and honesty guides every facet and step of empowerment evaluation. This commitment is considered a necessary but not sufficient factor. It sets the right tone in pursuit of quality; and the steps and facets help to ensure that this intent is translated into an honest appraisal.

STEPS OF EMPOWERMENT EVALUATION

There are three steps involved in helping others learn to evaluate their own programs: (a) developing a mission or unifying purpose; (b) taking stock or determining where the program stands, including strengths and weaknesses; and (c) planning for the future by establishing goals and helping participants determine their own strategies to accomplish program goals and objectives. In addition, empowerment evaluators help program staff members and participants determine the type of evidence required to document and monitor progress credibly toward their goals.

Mission

The first step in an empowerment evaluation is to ask program staff members and participants to define their mission. An empowerment evaluator facilitates an open session with as many staff members and participants as possible. They are asked to generate key phrases that capture the mission of the program or project. This is done even when an existing mission statement exists, because there are typically many new participants and the initial document may or may not have been generated in a democratic open forum. This allows fresh new ideas to become a part of the mission and it also allows participants an opportunity to voice their vision of the program. It is common for groups to learn how divergent their participants' views are about the program, even when working together for years. The evaluator records these phrases, typically on a poster sheet. Then a workshop participant is asked to volunteer to write these telescopic phrases into a paragraph or two. This document is shared with the group, revisions and corrections are made in the process, and then the group is asked to accept the document on a consensus basis – they don't have to be in favor of 100% of the document, they just have to be willing to live with it. The mission statement represents the values of the group and as such represents the foundation for the next step, taking stock.

Taking Stock

The second step in an empowerment evaluation is taking stock. It has two sections. The first involves generating a list of key activities that are crucial to the functioning of the program. Once again, the empowerment evaluator serves as a facilitator, asking program staff members and participants to list the most significant features and/or activities associated with the program. A list of 10 to 20 activities is sufficient. After generating this list, it is time to prioritize and determine which are the most important activities meriting evaluation at this time. One tool used to minimize the time associated with prioritizing activities involves voting with dots. The empowerment evaluator gives each participant 5 dots and asks them to place them by the activity they want to

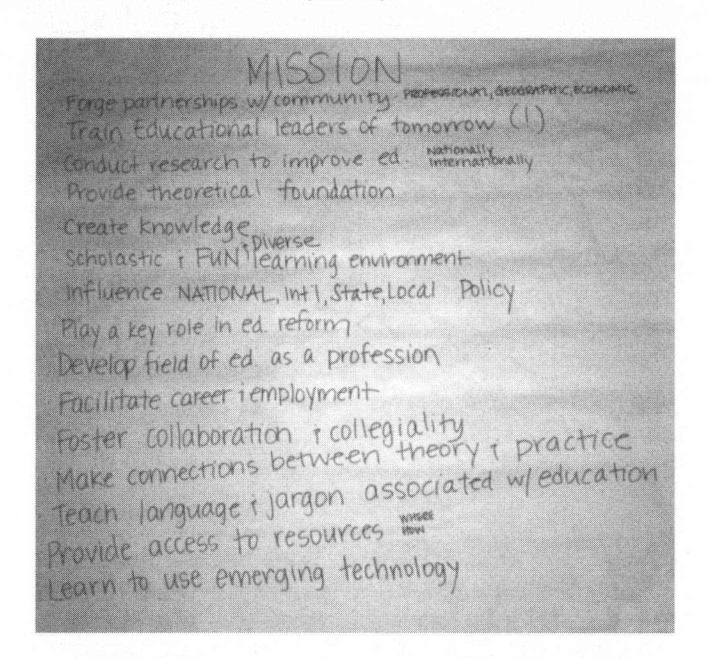

Fig. 1. This is a poster documenting the first phase in developing a School of Education Mission statement. The facilitator records program participant phrases concerning the mission on this poster sheet. Later these phrases are transformed into a more formal document, ranging from a paragraph to a couple of pages.

focus on. They can distribute them across 5 different activities or place all 5 on one activity. Counting the dots easily identifies the top 10 activities. The 10 activities with the most dots become the prioritized list of activities meriting evaluation at that time. (This process avoids long arguments about why one activity is valued more than another, when both activities are included in the review anyway.)

The second phase of taking stock involves rating the activities. Program staff members and participants are asked to rate each activity on a 1 to 10 scale, with 10 being the highest level and 1 being the lowest. Typically, they rate each of the activities while in their seats on their own piece of paper. Then they are asked to come up to the front of the room and record their ratings on a poster sheet of paper. This allows for some degree of independence in rating. In addition, it minimizes a long stream of second guessing, and checking to see what others are rating the same activities while recording ratings in the front of the room on the poster sheet.

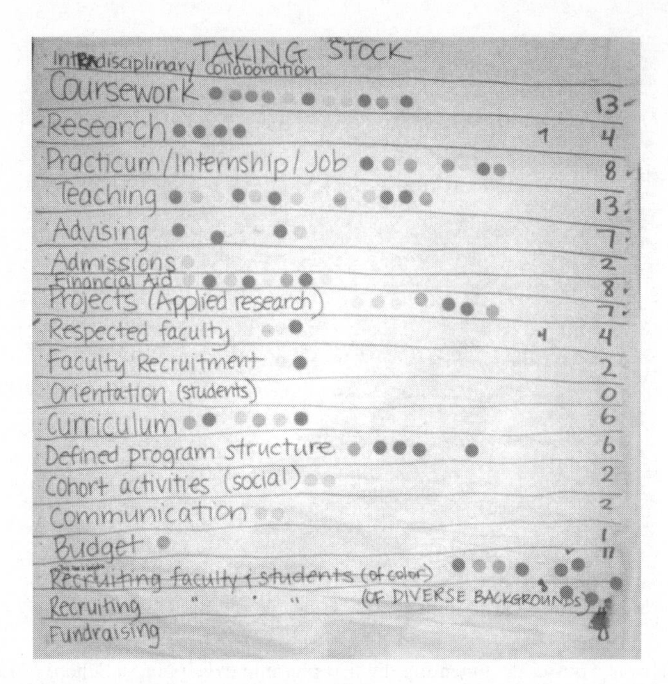

Fig. 2. This picture captures a typical Taking Stock prioritization exercise. Dots are used to vote for the most significant activities in the program. The total number of dots for each activity have been added on the right hand side of the poster. The activities with the most dots are selected for the second stage of the Taking Stock exercise – rating the activities.

At the same time, there is nothing confidential about the process. Program staff members and participants place their initials at the top of the matrix and then record their ratings for each activity. Contrary to most research designs, in general this system is designed to ensure that everyone knows and is influenced by each other's ratings (after recording them on the poster sheet). This is part of the socialization process that takes place in an empowerment evaluation, opening up the discussion and stepping toward more open disclosure – speaking one's truth.

The taking stock phase of an empowerment evaluation is conducted in an open setting for three reasons: (a) it creates a democratic flow of information and exchange of information, (b) it makes it more difficult for managers to retaliate because it is in an open forum, and (c) it increases the probability that the disclosures will be diplomatic because program staff members and participants must remain in that environment. Open discussions in a vacuum, without regard for workplace norms, are not productive. They are often unrealistic and can be counter-productive.

It is important that program staff members and participants be asked to begin by assessing program activities individually. They are more likely to give some activities low ratings if they are given an equal opportunity to speak positively about, or rate other, activities highly. In addition, they are more likely to give their program a higher rating if they are only asked to give an overall or gestalt rating on the program. The ratings can be totaled and averaged by person and by activity. This provides some insight into routinely optimistic and pessimistic participants. It allows participants to see where they stand in relation to their peers, which helps them calibrate their own assessments in the future. The more important rating, of course, is across the matrix by activity. Each activity receives a total and average. Combining the activity averages generates a total program rating, often lower than an external assessment rating. This represents the first baseline data concerning that specific program activity. This can be used to compare change over time.

All of this work sets the tone for one of the most important parts of the empowerment evaluation process – dialogue. The empowerment evaluator facilitates a discussion about the ratings. A survey would have accomplished the same task up to this point. However, the facilitator probes and asks why one person rated communication a 6 while two others rated it a 3 on the matrix. Participants are asked to explain their rating and provide evidence or documentation to support the rating. This plants the seeds for the next stage of empowerment evaluation – planning for the future, where they will need to specify the evidence they plan to use to document that their activities are helping them accomplish their goals. The empowerment evaluator serves as a critical friend during this stage, facilitating discussion and making sure everyone is heard and at the same time being critical and asking "what do you mean by that" or asking for additional clarification and substantiation about a particular rating or view. "Doing evaluation" in this manner helps program staff and participants internalize the drive to recognize and foster quality.

Participants are asked for both the positive and negative basis for their ratings. For example, if they give communication a 3, they are asked why a 3? The typical response is because there is poor communication and they proceed to list reasons for this problem. The empowerment evaluator listens and helps record the information and then asks the question again, focusing on why it was a 3 instead of a 1. In other words, there must be something positive to report as well. An important part of empowerment evaluation involves building on strengths and even in weak areas there is typically something positive going on that can be used to strengthen that activity or other activities. If the effort becomes exclusively problem focused then all anyone sees is problems instead of strengths and opportunities to build and improve on practice.

Some participants give their programs or specific activities unrealistically high ratings. The absence of appropriate documentation, peer ratings, and a reminder about the realities of their environment – such as a high drop-out rate, students bringing guns to school, and racial violence in a high school – help participants recalibrate their rating. However, in some cases, ratings stay higher than peers consider appropriate. The significance of this process, however, is not the actual rating so much as it is the creation of a baseline, as noted earlier, from which future progress can be measured. In addition, it sensitizes program participants to the necessity of collecting data to support assessments or appraisals.

After examining four or five examples, beginning with divergent ones and ending with similar ratings (to determine if there are totally different reasons for the same or similar ratings), this phase of the workshop is generally complete. The group or a designated subcommittee continues to discuss the ratings and the group is asked to return to the next workshop (planning for the future) with the final ratings and a brief description or explanation of what the ratings meant. (This is normally shared with the group for review and a consensus is sought concerning the document.) This process is superior to surveys because it generally has a higher response rate – toward 100%. In addition, it allows participants to discuss what they meant by their ratings, recalibrate and revise their ratings based on what they learn – minimizing "talking past each other" about certain issues or other miscommunications such as defining terms differently and using radically different rating systems. Participants learn what a 3 and an 8 mean to individuals in the group in the process of discussing and arguing about these ratings. This is a form of norming, helping create shared meanings and interpretations within a group. Norming is an important contribution to quality in any endeavor.

Planning for the Future

After rating their program's performance and providing documentation to support that rating, program participants are asked "where they want to go from here." How would they like to improve on what they do well and not so well? The empowerment evaluator asks the group to use the taking stock list of activities as the basis for their plans for the future – so that their mission guides their taking stock phase, and their taking stock phase shapes their planning for

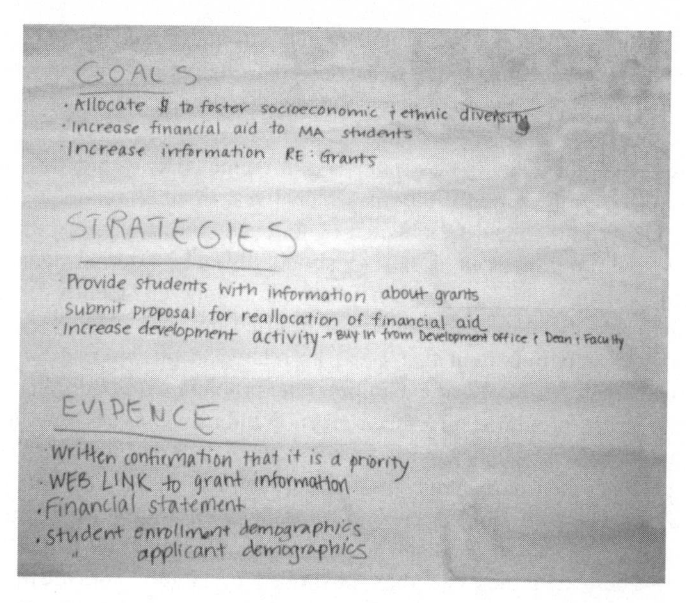

Fig. 3. This is a Planning for the Future poster focusing on Financial Aid. The three categories used to organize this stage of the process are: Goals, Strategies, and Evidence. The facilitator writes the participants' ideas down in the appropriate category on these poster sheets.

the future. This creates a thread of coherence and an audit trail for each step of their evaluation and action plans.

Program staff members and participants are asked to list their goals based on the results of their taking stock exercise. They set specific goals associated with each activity. In addition, the empowerment evaluator asks members of the group for strategies to accomplish each goal. In addition, they are asked to generate forms of evidence to monitor progress toward specified goals. Program staff members and participants supply all of this information.

The empowerment evaluator is not superior or inferior in the process. They are co-equals. They add ideas as deemed appropriate without dominating discussion. Their primary role is to serve as a coach, facilitator, and critical evaluative friend. They must be able to serve as a facilitator, helping program members and participants process and be heard. On the other hand, if that is all they do then nothing gets accomplished in terms of achieving specified goals. The evaluator must also be analytical and critical, asking or prompting participants to clarify, document, and evaluate what they are doing. If the evaluator is only critical and analytical, the group will walk away from the endeavor. The empowerment evaluator must maintain a balance of these talents or team up with other coaches (from within the group or outside the group) who can help them maintain this balance.

The selected goals should be established in conjunction with supervisors and clients to ensure relevance from both perspectives. In addition, goals should be realistic, taking into consideration such factors as initial conditions, motivation, resources, and program dynamics. They should also take external standards into consideration, e.g. accreditation agency standards, superintendent's 5-year plan, board of trustee dictates, board standards, and so on.

In addition, it is important that goals be related to the program's activities, talents, resources, and scope of capability. One problem with traditional external evaluation is that programs have been given grandiose goals or long-term goals that participants could only contribute to in some indirect manner. There was no link between their daily activities and ultimate long-term program outcomes in terms of these goals. In empowerment evaluation, program participants are encouraged to select intermediate goals that are directly linked to their daily activities. These activities can then be linked to larger, more diffuse goals, creating a clear chain of outcomes.

Program participants are encouraged to be creative in establishing their goals. A brainstorming approach is often used to generate a new set of goals. Individuals are asked to state what they think the program should be doing. The list generated from this activity is refined, reduced, and made realistic after the brainstorming phase, through a critical review and consensual agreement process.

There are also a bewildering number of goals to strive for at any given time. As a group begins to establish goals based on this initial review of their program, they realize quickly that a consensus is required to determine the most significant issues to focus on. These are chosen according to significance to the operation of the program, such as teaching; timing or urgency, such as recruitment or budget issues; and vision, including community building and learning processes.

Developing Strategies

Program participants are also responsible for selecting and developing strategies to accomplish program objectives. The same process of brainstorming, critical review, and consensual agreement is used to establish a set of strategies. These strategies are routinely reviewed to determine their effectiveness and appropriateness. Determining appropriate strategies, in consultation with sponsors and clients, is an essential part of the empowering process. Program participants are typically the most knowledgeable about their own jobs, and this approach acknowledges and uses that expertise – and in the process, puts them back in the "driver's seat."

Documenting Progress

Program staff members and participants are asked what type of documentation is required to monitor progress toward their goals. This is a critical step in the pursuit of quality. Each form of documentation is scrutinized for relevance to avoid devoting time to collecting information that will not be useful or relevant. Program participants are asked to explain how a given form of documentation is related to specific program goals. This review process is difficult and time-consuming but prevents wasted time and disillusionment at the end of the process. In addition, documentation must be credible and rigorous if it is to withstand the criticism that this evaluation is self-serving. (See Fetterman, 1994b, for additional discussion on this topic.)

The entire process of establishing a mission, taking stock, and planning for the future creates an implicit logic model or program theory, demonstrating how there is "nothing as practical as a good theory" of action (grounded in participants' own experiences). The facets of empowerment evaluation represent a meta-theory of the approach based on actual practice. The facets are discussed after presenting a brief case example to highlight the steps and the logic of this approach.

THE INSTITUTE: A CASE EXAMPLE

The California Institute of Integral Studies is an independent graduate school located in San Francisco. The Commission for Senior Colleges and Universities of Schools and Colleges (WASC) has accredited it since 1981. The accreditation process requires periodic self-evaluations. The Institute adopted an empowerment evaluation approach as a tool to institutionalize evaluation as part of the planning and management of operations and to respond to the accreditation self-study requirement. Empowerment evaluation is a form of self-evaluation that is designed to foster program improvement and self-determination (instead of dependency). It has three stages (as discussed above): the first focuses on identifying the program mission. The second, taking stock, examines and assesses the merit of a unit. The third is designed to establish goals and strategies to improve program practice. This third stage is part of strategic planning and is built on the collaborative foundation of understanding informed by mission and taking stock activities.

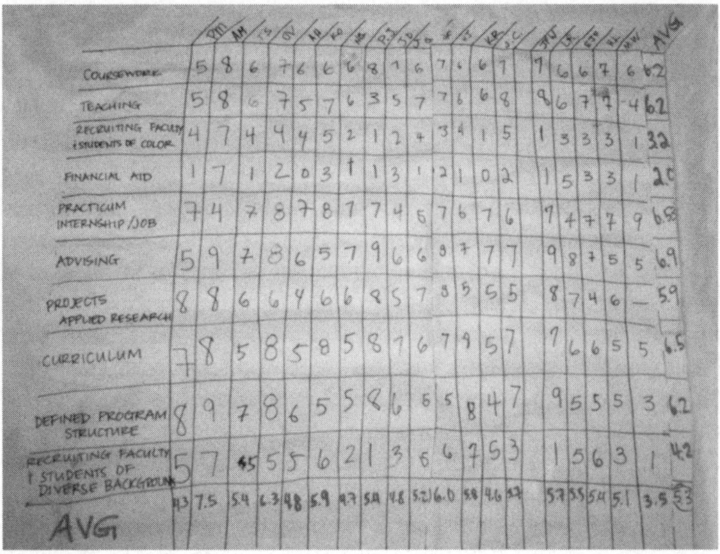

Fig. 4. This is a picture of the matrix used to facilitate this stage of the empowerment evaluation process. Activities are listed on the left column. Participant initials are on the top of the matrix. Individual ratings are listed for each activity in the column directly below the participant's initials. The averages are recorded on the bottom and on the right hand side of the spreadsheet. This worksheet provides a useful mechanism to enter into a dialogue about the status of the program.

All units in the Institute – including academic, governance, and administrative units – conducted self-evaluations. The purpose of these self-evaluations was to improve operations and build a base for planning and decision making. In addition to focusing on improvement, these self-evaluations contributed to institutional accountability.

Workshops were conducted throughout the Institute to provide training in evaluation techniques and procedures. All unit heads attended the training sessions held over 3 days. They served as facilitators in their own groups. Training and individual technical assistance was also provided throughout the year for governance and other administrative groups, including the Office of the President and the Development Office. (See Appendix for additional details.)

The self-evaluation process required thoughtful reflection and inquiry. The units described their purpose and listed approximately 10 key unit activities that characterized their unit. Members of the unit democratically determined the top 10 activities that merit consideration and evaluation. Then each member of a unit evaluated each activity by rating the activities on a 1 to 10 scale. Individual ratings were combined to produce a group or unit rating for each activity and one for the total unit. Unit members reviewed these ratings. A sample matrix is provided below to illustrate how this process was implemented.

Unit members discussed and dissected the meaning of the activities listed in the matrix and the ratings given to each activity. This exchange provided unit members with an opportunity to establish norms concerning the meaning of terms and ratings in an open and collegial atmosphere. Unit members were also required to provide evidence or documentation for each rating and/or to identify areas in which additional documentation was needed. These self-evaluations represented the first baseline data about program and unit operations concerning the entire Institute. This process was superior to survey methods for four reasons: (a) unit members determined what to focus on to improve their own programs – improving the validity of the effort and the buy-in required to implement recommendations; (b) a majority of members of the community were immersed in the evaluation experience, making the process of building a culture of evidence and a community of learners as important as the specific evaluative outcomes; (c) members of the community entered into a dialogue about the ratings, helping them establish norms and common understandings; and (c) there was a 100% return rate (as compared with typically low return rates for surveys).

These self-evaluations have already been used to implement specific improvements in program practice. This process has been used to place old problems in a new light, leading to solutions, adaptations, and new activities for the future. It has also been used to reframe existing data from traditional sources, enabling participants to give meaningful interpretation to data they already collect. In addition, self-evaluations have been used to ensure programmatic and academic accountability. For example, the Psychology program decided to discontinue its Ph.D. program as part of the self-evaluation process. This was a significant WASC (accreditation agency) and Institute concern of long standing. The core of the problem was that there weren't enough faculty in the program to properly serve their students. The empowerment evaluation process provided a vehicle for the faculty to come to terms with this problem in an open, self-conscious manner. They were dedicated to serving students properly but when they finally sat down and analyzed faculty-student ratios and faculty dissertation loads the problem became self-evident. (The faculty had complained about the workload and working conditions before, but they had never consciously analyzed, diagnosed, and documented this problem because they did not have the time or a simple, nonthreatening mechanism to assess themselves.) Empowerment evaluation provided Institute faculty with a tool to evaluate the program in light of scarce resources and made an executive decision to discontinue the program. Similarly, the all on-line segment of one of the Institute's Ph.D. programs has been administratively merged with a distance learning component of the same program as a result of this self-evaluative process. This was done to provide greater efficiencies of scale, improved monitoring and supervision, and more face-to-face contact with the Institute. (See Fetterman 1996b and 1996c for a description of one of these on-line educational programs.)

The last stage of an empowerment evaluation involves building plans for the future based on these evaluations. All units at the Institute completed their plans for the future, and these data were used to design an overall strategic plan. Goals and strategies were specified. In addition, relevant evidence was specified to monitor progress toward selected goals. This process ensures community involvement and commitment to the effort, generating a plan that is grounded in the reality of unit practice. The Provost institutionalized this process by requiring self-evaluations and unit plans on an annual basis to facilitate program improvement and contribute to institutional accountability.

FACETS OF EMPOWERMENT EVALUATION

In this new context, training, facilitation, advocacy, illumination, and liberation are all facets – if not developmental stages – of empowerment evaluation. Rather than additional roles for an evaluator whose primary function is to assess worth (as defined by Stufflebeam, 1994 and Scriven, 1967), these facets are an integral part of the evaluation process. Cronbach's developmental focus is relevant: the emphasis is on program development, improvement, and lifelong learning.

Training

In one facet of empowerment evaluation, evaluators teach people to conduct their own evaluations and thus become more self-sufficient. This approach desensitizes and demystifies evaluation and ideally helps organizations internalize evaluation principles and practices, making evaluation an integral part of program planning. Too often, an external evaluation is an exercise in dependency rather than an empowering experience: in these instances, the process ends when the evaluator departs, leaving participants without the knowledge or experience to continue for themselves. In contrast, an evaluation conducted by program participants is designed to be ongoing and internalized in the system, creating the opportunity for capacity building.

In empowerment evaluation, training is used to map out the terrain, highlighting categories and concerns. It is also used in making preliminary assessments of program components, while illustrating the need to establish goals, strategies to achieve goals, and documentation to indicate or substantiate progress. Training a group to conduct a self-evaluation can be considered equivalent to developing an evaluation or research design (as that is the core of the training), a standard part of any evaluation. This training is ongoing, as new skills are needed to respond to new levels of understanding. Training also becomes part of the self-reflective process of self-assessment (on a program level) in that participants must learn to recognize when more tools are required to continue and enhance the evaluation process. This self-assessment process is pervasive in an empowerment evaluation – built into every part of a program, even to the point of reflecting on how its own meetings are conducted and feeding that input into future practice.

In essence, empowerment evaluation is the "give someone a fish and you feed her for a day; teach her to fish, and she will feed herself for the rest of her life" concept, as applied to evaluation. The primary difference is that in empowerment evaluation the evaluator and the individuals benefiting from the evaluation are typically on an even plane, learning from each other.

Facilitation

Empowerment evaluators serve as coaches or facilitators to help others conduct a self-evaluation. In my role as coach, I provide general guidance and direction to the effort, attending sessions to monitor and facilitate as needed. It is critical to emphasize that the staff are in charge of their effort; otherwise, program participants initially tend to look to the empowerment evaluator as expert, which makes them dependent on an outside agent. In some instances, my task is to clear away obstacles and identify and clarify miscommunication patterns. I also participate in many meetings along with internal empowerment evaluators, providing explanations, suggestions, and advice at various junctures to help ensure that the process has a fair chance.

An empowerment evaluation coach can also provide useful information about how to create facilitation teams (balancing analytical and social skills), work with resistant (but interested) units, develop refresher sessions to energize tired units, and resolve various protocol issues. Simple suggestions along these lines can keep an effort from backfiring or being seriously derailed. A coach may also be asked to help create the evaluation design with minimal additional support.

Whatever her contribution, the empowerment evaluation coach must ensure that the evaluation remains in the hands of program personnel. The coach's task is to provide useful information, based on her evaluator's training and past experience, to keep the effort on course and in pursuit of quality at every stage.

Advocacy

A common workplace practice provides a familiar illustration of self-evaluation and its link to advocacy on an individual level. Employees often collaborate with both supervisor and clients to establish goals, strategies for achieving those goals and documenting progress, and realistic timelines. Employees collect data on their own performance and present their case for their performance appraisal. Self-evaluation thus becomes a tool of advocacy. This individual self-evaluation process is easily transferable to the group or program level.

Illumination

Illumination is an eye-opening, revealing, and enlightening experience. Typically, a new insight or understanding about roles, structures, and program dynamics is developed in the process of determining worth and striving for program improvement (see Parlett & Hamilton, 1976). Empowerment evaluation is illuminating on a number of levels. For example, an administrator in one empowerment evaluation, with little or no research background, developed a testable, researchable hypothesis in the middle of a discussion about indicators and self-evaluation. It was not only illuminating to the group (and to her), it revealed what they could do as a group when given the opportunity to think about problems and come up with workable options, hypotheses, and tests. This experience of illumination holds the same intellectual intoxication each of us experienced the first time we came up with a researchable question. The process creates a dynamic community of learners as people engage in the art and science of evaluating themselves. The culture of evaluation and evidence, created by this community of learners, is a fundamental tool to help internalize the pursuit of quality and improvement.

Liberation

Illumination often sets the stage for liberation. It can unleash powerful, emancipatory forces for self-determination. Liberation is the act of being freed or freeing oneself from preexisting roles and constraints. It often involves new conceptualizations of oneself and others. Empowerment evaluation can also be liberating. Many of the examples in this discussion demonstrate how helping individuals take charge of their lives – and find useful ways to evaluate themselves – liberates them from traditional expectations and roles. They also demonstrate how empowerment evaluation enables participants to find new opportunities, see existing resources in a new light, and redefine their identities and future roles. The ability to reframe a problem or a situation and break out of the mold is often what "good" evaluation can contribute to a program. It is often referred to as value or quality added. However, in this case, program staff members and participants learn how to use evaluation to break away from the mold and their pre-existing roles.

CAVEATS AND CONCERNS: THE QUALITY OF THE PROCESS

Is Research Rigor Maintained?

This case study presented a picture of how research and evaluation rigor is maintained. Mechanisms employed to maintain rigor included: workshops and training, democratic participation in the evaluation (ensuring that majority and minority views are represented), quantifiable rating matrices to create a baseline to measure progress, discussion and definition of terms and ratings (norming), scrutinizing documentation, and questioning findings and recommendations. These mechanisms help ensure that program participants are critical, analytical, and honest. These mechanisms speak to the issue of internalizing quality as a basic value in the group.

Empowerment evaluation is one approach among many being used to address social, educational, industrial, health care, and many other problems. As with the exploration and development of any new frontier, this approach requires adaptations, alterations, and innovations. This does not mean that significant compromises must be made in the rigor required to conduct evaluations. Although I am a major proponent of individuals taking evaluation into their own hands and conducting self-evaluations, I recognize the need for adequate research, preparation, and planning. These first discussions need to be supplemented with reports, texts, workshops, classroom instruction, and apprenticeship experiences if possible. Program personnel new to evaluation should seek the assistance of an evaluator to act as coach, assisting in the design and execution of an evaluation. Further, an evaluator must be judicious in determining when it is appropriate to function as an empowerment evaluator or in any other evaluative role.

Does This Abolish Traditional Evaluation?

New approaches require a balanced assessment. A strict constructionist perspective may strangle a young enterprise; too liberal a stance is certain to transform a novel tool into another fad. Colleagues who fear that we are giving evaluation away are right in one respect – we are sharing it with a much broader population. But those who fear that we are educating ourselves out of a job are only partially correct. Like any tool, empowerment evaluation is designed to address a specific evaluative need. It is not a substitute for other forms of evaluative inquiry or appraisal. We are educating others to manage their own affairs in areas they know (or should know) better than we do. At the same time, we are creating new roles for evaluators to help others help themselves.

How Objective Can a Self-Evaluation Be?

Objectivity is a relevant concern. We needn't belabor the obvious point that science and specifically evaluation have never been neutral. Anyone who has had to roll up her sleeves and get her hands dirty in program evaluation or policy arenas is aware that evaluation, like any other dimension of life, is political, social, cultural, and economic. It rarely produces a single truth or conclusion. In the context of a discussion about self-referent evaluation, Stufflebeam (1994) states:

> As a practical example of this, in the coming years U.S. teachers will have the opportunity to have their competence and effectiveness examined against the standards of the National Board for Professional Teaching Standards and if they pass to become nationally certified (p. 331).

Regardless of one's position on this issue, evaluation in this context is a political act. What Stufflebeam considers an opportunity, some teachers consider a threat to their livelihood, status, and role in the community. This can be a screening device in which social class, race, and ethnicity are significant variables. The goal is "improvement," but the questions of for whom and at what price remain valid. Evaluation in this context or any other is not neutral – it is for one group a force of social change, for another a tool to reinforce the status quo.

According to Stufflebeam (1994), "objectivist evaluations are based on the theory that moral good is objective and independent of personal or merely human feelings. They are firmly grounded in ethical principles, strictly control bias or prejudice in seeking determinations of merit and worth." To assume that evaluation is all in the name of science or that it is separate, above politics, or "mere human feelings" – indeed, that evaluation is objective – is to deceive oneself and to do an injustice to others. Objectivity functions along a continuum – it is not an absolute or dichotomous condition of all or none. Fortunately, such objectivity is not essential to being critical. For example, I support programs designed to help dropouts pursue their education and prepare for a career; however, I am highly critical of program implementation efforts. If the program is operating poorly, it is doing a disservice both to former dropouts and to taxpayers.

One needs only to scratch the surface of the "objective" world to see that values, interpretations, and culture shape it. Whose ethical principles are evaluators grounded in? Do we all come from the same cultural, religious, or even academic tradition? Such an ethnocentric assumption or assertion flies in the face of our accumulated knowledge about social systems and evaluation. Similarly, assuming that we can "strictly control bias or prejudice" is naive, given the wealth of literature available on the subject, ranging from discussions about cultural interpretation to reactivity in experimental design.

What About Participant or Program Bias?

The process of conducting an empowerment evaluation requires the appropriate involvement of stakeholders. The entire group – not a single individual, not the external evaluator or an internal manager – is responsible for conducting the evaluation. The group can thus serve as a check on individual members, moderating their various biases and agendas.

No individual operates in a vacuum. Everyone is accountable in one fashion or another and thus has an interest or agenda to protect. A school district may have a 5-year plan designed by the superintendent; a graduate school may have to satisfy requirements of an accreditation association; an outside evaluator may have an important but demanding sponsor pushing either timelines or results, or may be influenced by training to use one theoretical approach rather than another.

In a sense, empowerment evaluation minimizes the effect of these biases by making them an explicit part of the process. The example of a self-evaluation in a performance appraisal is useful again here. An employee negotiates with his or her supervisor about job goals, strategies for accomplishing them, documentation of progress, and even the timeline. In turn, the employee works with clients to come to an agreement about acceptable goals, strategies, documentation, and timelines. All of this activity takes place within corporate, institutional, and/or community goals, objectives, and aspirations. The larger context, like theory, provides a lens in which to design a self-evaluation. Self-serving forms of documentation do not easily persuade supervisors and clients.

Once an employee loses credibility with a supervisor, it is difficult to regain it. The employee thus has a vested interest in providing authentic and credible documentation. Credible data (as agreed on by supervisor and client in negotiation with the employee) serve both the employee and the supervisor during the performance appraisal process.

Applying this approach to the program or community level, superintendents, accreditation agencies, and other "clients" require credible data. Participants in an empowerment evaluation thus negotiate goals, strategies, documentation, and timelines. Credible data can be used to advocate for program expansion, redesign, and/or improvement. This process is an open one, placing a check on self-serving reports. It provides an infrastructure and network to combat institutional injustices. It is a highly (often brutally) self-critical process. Program staff members and participants are typically more critical of their own program than an external evaluator, often because they are more familiar with their program and would like to see it serve its purpose(s) more effectively.[4]

Empowerment evaluation is successful because it adapts and responds to existing decision making and authority structures on their own terms (see Fetterman, 1993). It also provides an opportunity and a forum to challenge authority and managerial facades by providing data about actual program operations – from the ground up. This approach is particularly valuable for disenfranchised people and programs to ensure that their voices are heard and that real problems are addressed.

POSITIONS OF PRIVILEGE

Empowerment evaluation is grounded in my work with the most marginalized and disenfranchised populations, ranging from urban school systems to community health programs in South African townships, who have educated me about what is possible in communities overwhelmed by violence, poverty, disease, and neglect. They have also repeatedly sensitized me to the power of positions of privilege. One dominant group has the vision, makes and changes the rules, enforces the standards, and need never question its own position or seriously consider any other. In such a view, differences become deficits rather than additive elements of culture. People in positions of privilege dismiss the contributions of a multicultural world. They create rational policies and procedures that systematically deny full participation in their community to people who think and behave differently.

Evaluators cannot afford to be unreflective about the culturally embedded nature of our profession. There are many tacit prejudgments and omissions embedded in our primarily Western thought and behavior. These values, often assumed to be superior, are considered natural. Western philosophies, however, have privileged their own traditions and used them to judge others who may not share them, disparaging such factors as ethnicity and gender. In addition, they systematically exclude other ways of knowing. Some evaluators are convinced that there is only one position and one sacred text in evaluation, justifying exclusion or excommunication for any "violations" or wrong thinking (see Stufflebeam, 1994). Scriven's (1991, p. 260) discussion about perspectival evaluation is instructive in this context, highlighting the significance of adopting multiple perspectives and including new perspectives.

We need to keep open minds, including alternative ways of knowing – but not empty heads. Skepticism is healthy; cynicism, blindness, and condemnation are not, particularly for emerging evaluative forms and adaptations. New approaches in evaluation and even new ways of knowing are needed if we are to expand our knowledge base and respond to pressing needs. As Campbell (1994) states, we should not "reject the new epistemologies out of hand. . . . Any specific challenge to an unexamined presumption of ours should be taken seriously" (p. 293). Patton (1994) might be right "that the world will not end in a subjective bang, but in a boring whimper as voices of objectivity [drift] off into the chaos" (p. 312).

Evaluation must change and adapt as the environment changes, or it will either be overshadowed by new developments or – as a result of its unresponsiveness and irrelevance – follow the path of the dinosaurs to extinction. People are demanding much more of evaluation and are not tolerant of the limited role of the outside expert who has no knowledge of or vested interest in their program or community. Participation, collaboration, and empowerment are becoming requirements in many community-based evaluations, not recommendations. Program participants are conducting empowerment and other forms of self – or participatory evaluations with or without us (the evaluation community). I think it is healthier for all parties concerned to work together to improve practice rather than ignore, dismiss, and condemn evaluation practice; otherwise, we foster the development of separate worlds operating and unfolding in isolation from each other.

DYNAMIC COMMUNITY OF LEARNERS

Many elements must be in place for empowerment evaluation to be effective and credible. Participants must have the latitude to experiment, taking both risks and responsibility for their actions. An environment conducive to sharing successes and failures is also essential. In addition, an honest, self-critical, trusting, and supportive atmosphere is required. Conditions need not be perfect to initiate this process. However, the accuracy and usefulness of self-ratings improve dramatically in this context. An outside evaluator who is charged with monitoring the process can help keep the effort credible, useful, and on track, providing additional rigor, reality checks, and quality controls throughout the evaluation. Without any of these elements in place, the exercise may be of limited utility and potentially self-serving. With many of these elements in place, the exercise can create a dynamic community of transformative learning.

CONCLUSION

Empowerment evaluation is fundamentally a democratic process. The entire group – not a single individual, not the external evaluator or an internal manager – is responsible for conducting the evaluation. The group can thus serve as a check on its own members, moderating the various biases and agendas of individual members. The evaluator is a co-equal in this endeavor, not a superior and not a servant. As a critical friend, the evaluator can question shared biases or "group think."

As is the case in traditional evaluation, everyone is accountable in one fashion or another and thus has an interest or agenda to protect. A school district may have a five-year plan designed by the superintendent; a graduate school may have to satisfy requirements of an accreditation association; an outside evaluator may have an important but demanding sponsor pushing either timelines or results, or may be influenced by training to use one theoretical approach rather than another. Empowerment evaluations, like all other evaluations, exist within a context. However, the range of intermediate objectives linking what most people do in their daily routine and macro goals is almost infinite. People often feel empowered and self-determined when they can select meaningful intermediate objectives that are linked to larger, global goals.

Despite its focus on self-determination and collaboration, empowerment evaluation and traditional external evaluation are not mutually exclusive – to the contrary, they enhance each other. In fact, the empowerment evaluation process produces a rich data source that enables a more complete external examination. In the empowerment evaluation design developed in response to the school's accreditation self-study requirement presented in this discussion, a series of external evaluations were planned to build on and enhance self-evaluation efforts. A series of external teams were invited to review specific programs. They determined the evaluation agenda in conjunction with department faculty, staff, and students. However, they operated as critical friends providing a strategic consultation rather than a compliance or traditional accountability review. Participants agreed on the value of an external perspective to add insights into program operation, serve as an additional quality control, sharpen inquiry, and improve program practice. External evaluators can also help determine the merit and worth of various activities. An external evaluation is not a requirement of empowerment evaluation, but it is certainly not mutually exclusive. Greater coordination between the needs of the internal and external forms of evaluation can provide a reality check concerning external needs and expectations for insiders, and a rich database for external evaluators. The pursuit of quality can become a rich and rewarding enterprise when internal and external forces unite for a common cause.

The external evaluator's role and productivity is also enhanced by the presence of an empowerment or internal evaluation process. Most evaluators operate significantly below their capacity in an evaluation because the program lacks even rudimentary evaluation mechanisms and processes. The external evaluator routinely devotes time to the development and maintenance of elementary evaluation systems. Programs that already have a basic self-evaluation process in place enable external evaluators to begin operating at a much more sophisticated level.

A matrix or similar design to further systematize internal evaluation activity facilitates comparison and analysis on a larger scale. Another approach involves a more artistic approach using green and red dots to signify progress or deterioration concerning specific topics of concern. The dots have a strong visual impact and can be quantified. Any system can work if it provides participants with straightforward, user-friendly tools to make credible judgments about where they are at any given point in time; provide consistent patterns (including those with a strong visual impact) that are meaningful to them; facilitate comparison – across individuals, categories, and programs; and stimulate constructive activity to improve program practice.

Finally, it is hoped that empowerment evaluation will benefit from the artful shaping of our combined contributions rather than follow any single approach or strategy. As Cronbach (1980) urged over two decades ago: "It is better for an evaluative inquiry to launch a small fleet of studies than to put all its resources into a single approach" (p. 7).

NOTES

1. It has been institutionalized as part of the Collaborative, Participatory, and Empowerment Evaluation Topical Interest Group (TIG). David Fetterman is the current TIG chair. All interested evaluators are invited to join the TIG and attend our business meetings, which are open to any member of the association.

2. Although there are many problems with the standards and the application of the standards to empowerment evaluation, and the fact that they have not been formally adopted by any professional organization, they represent a useful tool for self-reflection and examination. Empowerment evaluation meets or exceeds the spirit of the standards in terms of utility, feasibility, propriety, and accuracy (see Fetterman, 1996, 2000 for a detailed examination).

3. See Stevenson, Mitchell, and Florin (1995) for a detailed explanation about the distinctions concerning levels of organizations. See also Zimmerman (in press) for more detail about empowerment theory focusing on psychological, organizational, and community levels of analysis.

4. There are many useful mechanisms to enhance a self-critical mode. Beginning with an overall assessment of the program often leads to inflated ratings. However, asking program participants to assess program components before asking for an overall assessment facilitates the self-critical process. In addition, allowing individuals to comment on successful parts of the program typically enables them to comment openly on problematic components.

REFERENCES

Bandura, A. (1982). Self-efficacy mechanism in human agency. *American Psychologist*, *37*, 122–147.

Campbell, D. T. (1994). Retrospective and prospective on program impact assessment. *Evaluation Practice*, *15*(3), 291–298.

Choudhary, A., & Tandon, R. (1988). *Participatory evaluation*. New Delhi, India: Society for Participatory Research in Asia.

Cronbach, L. J. (1980). *Toward reform of program evaluation*. San Francisco: Jossey-Bass.

Dunst, C. J., Trivette, C. M., & LaPointe, N. (1992). Toward clarification of the meaning and key elements of empowerment. *Family Science Review*, *5*(1–2), 111–130.

Fetterman, D. M. (1993). *Speaking the language of power: Communication, collaboration, and advocacy (Translating ethnography into action)*. London: Falmer.

Fetterman, D. M. (1994a). Empowerment evaluation. Presidential address. *Evaluation Practice*, *15*(1), 1–15.

Fetterman, D. M. (1994b). Steps of empowerment evaluation: From California to Cape Town. *Evaluation and Program Planning*, *17*(3), 305–313.

Fetterman, D. M. (1995). In response to Dr. Daniel Stufflebeam's: Empowerment evaluation, objectivist evaluation, and evaluation standards: Where the future of evaluation should not go and where it needs to go. *Evaluation Practice*, *16*(2), 179–199.

Fetterman, D. M. (1996a). Empowerment evaluation: An introduction to theory and practice. In: D. M. Fetterman, S. Kaftarian & A. Wandersman (Eds), *Empowerment Evaluation: Knowledge and Tools for Self-Assessment and Accountability* (pp. 3–46). Thousand Oaks, CA: Sage.

Fetterman, D. M. (1996b). Ethnography in the virtual classroom. *Practicing Anthropology*, *18*(3), 2, 36–39.

Fetterman, D. M. (1996c). Videoconferencing: Enhancing communication on the Internet. *Educational Researcher*, *25*(4), 13–17.

Fetterman, D. M. (2000). *Foundations of Empowerment evaluation*. Thousand Oaks, CA: Sage.

Fetterman, D. M., & Haertel, E. H. (1990). *A school-based evaluation model for accelerating the education of students at-risk.* Clearinghouse on Urban Education. (ERIC Document Reproduction Service No. ED 313 495)

Fetterman, D. M., Kaftarian, S., & Wandersman, A. (1996). *Empowerment evaluation: Knowledge and tools for self-assessment and accountability.* Thousand Oaks, CA: Sage.

Joint Committee on Standards for Educational Evaluation (1994). *The program evaluation standards.* Thousand Oaks, CA: Sage.

Levin, H. M. (1996). Empowerment evaluation and accelerated schools. In: D. M. Fetterman, S. Kaftarian & A. Wandersman (Eds), *Empowerment Evaluation: Knowledge and Tools for Self-Assessment and Accountability* (pp. 49–64). Thousand Oaks, CA: Sage.

Mithaug, D. E. (1991). *Self-determined kids: Raising satisfied and successful children.* New York: Macmillan.

Mithaug, D. E. (1993). *Self-regulation therapy: How optimal adjustment maximizes gain.* New York: Praeger.

Oja, S. N., & Smulyan, L. (1989). *Collaborative action research.* Philadelphia: Falmer.

Papineau, D., & Kiely, M. C. (1994). Participatory evaluation: Empowering stakeholders in a community economic development organization. *Community Psychologist, 27*(2), 56–57.

Parlett, M., & Hamilton, D. (1976). Evaluation as illumination: A new approach to the study of innovatory programmes. In: D. Hamilton (Ed.), *Beyond the Numbers Game.* London: Macmillan.

Patton, M. (1994). Developmental evaluation. *Evaluation Practice, 15*(3), 311–319.

Rappaport, J. (1987). Terms of empowerment/exemplars of prevention: Toward a theory for community psychology. *American Journal of Community Psychology, 15,* 121–148.

Reason, P. (Ed.). (1988). *Human inquiry in action: Developments in new paradigm research.* Newbury Park, CA: Sage.

Sanders, J. R., Barley, Z. A., & Jenness, M. R. (1990). *Annual report: Cluster evaluation in science education.* Unpublished report.

Scriven, M. S. (1967). The methodology of evaluation. In: R. E. Stake (Ed.), *Curriculum Evaluation* (AERA Monograph Series on Curriculum Evaluation, Vol. 1). Chicago: Rand McNally.

Scriven, M. S. (1991). *Evaluation thesaurus* (4th ed.). Newbury Park, CA: Sage.

Shapiro, J. P. (1988). Participatory evaluation: Toward a transformation of assessment for women's studies programs and projects. *Educational Evaluation and Policy Analysis, 10*(3), 191–199.

Soffer, E. (1995). The principal as action researcher: A study of disciplinary practice. In: S. E. Noffke & R. B. Stevenson (Eds), *Educational Action Research: Becoming Practically Critical.* New York: Teachers College Press.

Stanford University and American Institutes for Research. (1992). *A design for systematic support for accelerated schools: In response to the New American Schools Development Corporation RFP for designs for a new generation of American schools.* Palo Alto, CA: Author.

Stevenson, J. F., Mitchell, R. E., & Florin, P. (1995). Evaluation and self-direction in community prevention coalitions. In: D. M. Fetterman, S. Kaftarian & A. Wandersman (Eds), *Empowerment Evaluation: Knowledge and Tools for Self-Assessment and Accountability* (pp. 208–233). Thousand Oaks, CA: Sage.

Stufflebeam, D. L. (1994). Empowerment evaluation, objectivist evaluation, and evaluation standards: Where the future of evaluation should not go and where it needs to go. *Evaluation Practice, 15*(3), 321–338.

Stull, D., & Schensul, J. (1987). *Collaborative research and social change: Applied anthropology in action.* Boulder, CO: Westview.

Tax, S. (1958). The Fox Project. *Human Organization, 17,* 17–19.

Whitmore, E. (1990). Empowerment in program evaluation: A case example. *Canadian Social Work Review*, *7*(2), 215–229.

Whyte, W. F. (Ed.) (1990). *Participatory action research.* Newbury Park, CA: Sage.

W. K. Kellogg Foundation (1992). *Transitions.* Battle Creek, MI: Author.

Zimmerman, M. A. (in press). Empowerment theory: Psychological, organizational, and community levels of analysis. In: J. Rappaport & E. Seldman (Eds), *Handbook of Community Psychology*. New York: Plenum.

Zimmerman, M. A., Israel, B. A., Schulz, A., & Checkoway, B. (1992). Further explorations in empowerment theory: An empirical analysis of psychological empowerment. *American Journal of Community Psychology*, *20*(6), 707–727.

Zimmerman, M. A., & Rappaport, J. (1988). Citizen participation, perceived control, and psychological empowerment. *American Journal of Community Psychology*, *16*(5), 725–750.

APPENDIX

California Institute of Integral Studies

1994–95 Unit Self-Evaluation Workshops for Unit Heads

Times and Places:
February 7, 4–6, 4th floor Conference Room
February 9, 4–6, 4th floor Conference Room
February 12, 12–2, All Saint's Church

Workshop facilitators:
David Fetterman, Karine Schomer, Mary Curran

Agenda

a. Introduction:
   i. Purpose of Unit Self-Evaluation: How it will feed into Academic Program Review, the WASC Self-Study,
   ii. Unit-Based Strategic Planning
b. Timelines & Deadlines, Report Formats
c. The Self-Empowerment Evaluation Method: Purpose, Process, Product
d. Overarching Institutional WASC Issues and Themes
e. Conducting a Demonstration: "Taking Stock" Session With a Volunteer Unit

Part I

a. Volunteer unit members describe their unit, its mission or purpose, and its relationship to the Institute's mission.

b. They list the unit's key activities.
c. They rate the quality and/or effectiveness of the top 10 key activities.
d. They list documentation/evidence to support the ratings.
e. They do an overall rating of the unit.

Part II: Break-Out Period

a. Volunteer unit members discuss among themselves their ratings of key activities: why did each person rate as he/she did? What did the ratings mean to each person? How did the unit achieve the rating given? How could it achieve a higher rating? Should ratings be adjusted in view of the discussion?
b. Meanwhile, other workshop participants form small groups. Each person writes a description of their unit, lists 5 key activities, rates these activities and the unit as a whole, and lists supporting documentation. Then all this is reported and discussed in the group.
c. Small group discussions are shared in large group.

Part III

a. Volunteer group adjusts its ratings of unit's key activities.
b. They adjust their overall rating of the unit.
c. They prioritize the key activities they have rated.
d. They list 2–3 preliminary recommendations for future courses of action: major goals, objectives related to the goals, strategies to achieve the objectives, documentation/evidence that would demonstrate success.

6. Instructions for Unit Tasks After Completion of the "Taking Stock" Session

a. Unit head writes up preliminary Unit Self-Evaluation Report based on the "Taking Stock" session, including unit mission/purpose, overall unit rating, rating of key activities, prioritization of key activities, list of documentation/evidence to support ratings, and preliminary recommendations for future courses of action.
b. Mini-sessions of the unit to review the report, discuss and adjust ratings, and build consensus about what the ratings mean and what everyone thinks about the unit.
c. Gathering and analyzing supporting documentation.
d. Unit head completes and submits final draft of the Unit Self Evaluation Report and supporting documentation.

1994–95 unit Self-Evaluation Report

Note 1: Form is available on disk from the Office of the Provost. (Please supply your own disk.)

Note 2: A single report format is being used for academic, administrative and governance units. Some items may therefore not be applicable to all units. They should be marked N/A.

Note 3: Reports and supporting documentation should be reviewed by person whom unit head reports to before submission to Office of the Provost.

Note 4: Reports should be distributed as widely as possible within unit and to other relevant units.

Name of Unit:
What larger unit does it belong to?
Academic/Administrative/Governance (circle one)
Name and title of unit head:
To whom does unit head report? (Title):

Part I – Unit Description

1. Mission or purpose of unit (narrative)
2. Relationship to Institute mission
3. Organizational structure of unit (narrative)
4. List of Ten (10) Key Activities performed by unit, prioritized (from "Taking Stock" session)
5. Other ongoing activities of unit (narrative)
6. Special projects of unit (list)
7. Direction of unit development over past three years (narrative)
8. Current three (3) major goals. Objectives related to goals, strategies being used to achieve objectives, and documentation/evidence that demonstrates success (narrative).
9. Number and names of core faculty by% time
10. Number and names of adjunct faculty
11. Number and names of staff, by title and% time
12. Number of students (persons over three years 1992–93 1993–94 1994–95 (est.)
13. Number of students (FTE) over past three years 1992–93 1993–94 1994–95 (est.)
14. Number of class enrollments over three years 1992–93 1993–94 1994–95 (est.)

15. Operational budget for past two years and current year 1992–93 1993–94 1994–95 (rev.)

Revenue Expense
(*Note*: Please footnote any figures requiring significant explanation.)

16. Institutional issues and themes of particular relevance to unit (refer to Strategic Directions)
17. WASC issues and themes of particular relevance to unit (refer to WASC recommendations, WASC Standards, WASC Self-Study Themes)

Part II – UNIT SELF-EVALUATION

1. Names of participants in Unit Self-Evaluation process (F-faculty, S-student, AS-administrative staff)
2. Date(s) of "Taking Stock" session
3. Dates and purpose of follow-up "mini-sessions"
4. Narrative of self-evaluation process
5. Overall rating of unit:
   Range Average
6. Rating of prioritized Ten (10) Key Activities (list)

Item	Range Average
(1)	
(2)	
(3)	
(4)	
(5)	
(6)	
(7)	
(8)	
(9)	
(10)	

7. List of documentation/evidence used to support ratings (attach documentation in appendix)
8. Discussion of findings (narrative based on overall ratings and ratings of prioritized Ten (10) Key Activities). Describe the Key Activities and how they are related to the mission or purpose of the unit. Explain what the individual and overall ratings mean. Explain the relevance of the documentation used. Summarize the overall strengths and weaknesses of the unit and progress made over the past three years.

9. Preliminary recommendations for 2–3 future courses of actions: major new goals, objectives related to the goals, strategies to achieve the objectives and documentation/evidence that would demonstrate success.
10. Evaluation and feedback on the Unit Self-Evaluation process

8. COMMUNICATING QUALITY AND QUALITIES: THE ROLE OF THE EVALUATOR AS CRITICAL FRIEND

Sharon F. Rallis and Gretchen B. Rossman

INTRODUCTION

Evaluators suggest different, sometimes quite dramatically different, meanings of the concept quality when discussing a program or service they have evaluated. Some judge the program's essential merit, goodness, or success; others assess the program's worth, often in light of the expenditure of public monies to support the program; and yet others describe its essential attributes or characteristics. At times, evaluators are not clear about which definition they rely upon, further confusing the conversation. For example, evaluations of various programs can lead to their cancellation because descriptions of attributes are misconstrued as judgments of worth, as in the case of a CPR training program that lost its funding largely because the formative evaluation identified certain weaknesses that led funders to assume the program was not valuable. The Oxford English Dictionary (OED, 1993) notes that, originally, the term quality meant "the nature or kind of something. Now, [it means] the relative nature or standard of something; the degree of excellence . . . possessed by a thing" (p. 2438). This evolution of the subtle meaning of the term raises the need to address explicitly the meaning of the concept of quality in evaluation.

We begin this chapter with a discussion of the definitional complexities of the concept of *quality*. We then describe the role of the evaluator as the critical friend of program leaders and participants, arguing that the major facets of

Visions of Quality, Volume 7, pages 107–120.
2001 by Elsevier Science Ltd.
ISBN: 0-7623-0771-4

quality are communicated through the dialogue facilitated by this role. In the final section, we offer ways for the evaluator to communicate the quality of the program or service through dialogic, narrative-textual, and visual-symbolic representations. The chapter ends with a call for multiple ways to represent the quality of the program or service, given the complexity of social programs and interventions.

MERIT, WORTH, AND ATTRIBUTES

We claim that the term quality has three overlapping yet distinct meanings. To support this claim, we analyze the term, drawing on current literature that presents various, intriguing definitions. The first meaning is traditionally associated with evaluation and describes the work of the evaluator as making a determination about the **merit** of a program or intervention – its quality. The Oxford English Dictionary (1993) defines merit as "a point of intrinsic quality; a commendable quality, an excellence, a good point" (p. 1748). Thus this facet of the concept focuses on "the degree of excellence . . . possessed by a thing" (p. 2438). Scriven (1991) defines merit as "'the intrinsic' value of evaluands, as opposed to extrinsic or system-related value/worth" (p. 227); he does not, however, explicate precisely on what basis that intrinsic value is to be assessed. This meaning of the concept *quality* suggests that the evaluator's role is to make determinations about the goodness or excellence of a program or service with the implication that judgments are also made about its shortcomings or weaknesses. For example, a large urban school system's study of their inclusion initiative focused on the merit of the effort: How is it working? What is working? What is not working? How can we improve the program? The scope of this judgment is on the program or service itself – a self-contained judgment – and is typical of the determinations evaluators are contracted to provide.

Worth, the second definition of the concept quality, moves beyond the program itself, focusing on the extrinsic value of the program or service: how it is assessed and valued by others. The urban school committee considered the *worth* of the inclusion initiative when it considered if it wanted to continue to expend monies (or to spend further monies) on the program: are the program activities and services valuable enough to students and the school to invest school funds? Pursuing this definition, Scriven (1991) notes that worth "usually refers to the value to an institution or collective [program or service], by contrast with intrinsic value, value by professional standards, or value to an individual consumer (merit)" (p. 382). He goes on to note that, "when people ask 'what something is worth', they are usually referring to *market value* and market value is a function of the market's (hypothesized) behavior toward it, not of the thing's intrinsic virtues or its virtue for an identifiable individual" (pp. 382–383, emphasis in original). This meaning is also central to evaluation work; evaluators express worth when they make judgments about the overall value of the program to some group of constituents or to society generally, including recommendations about funding.

Both of these judgments – merit and worth – depend, we argue, on a third definition, that is, detailed descriptions of the characteristics and **attributes** of complex social programs – their *qualities*. This facet of the concept *quality* focuses on the subtleties and nuances intrinsic to programs and services. The attributes – or qualities – are found in the "thick descriptions" (Geertz, 1973) that allow the evaluator and program personnel to interpret activities and events. From these thick descriptions and interpretations, they make judgments of merit or worth. The evaluator's role is to describe the "attribute, a property, a special feature or characteristic" (OED, 1993, p. 2438) of a program, providing the grounding for making judgments about merit and worth.

Integrating the first and third facets of the concept – judging merit and describing attributes – is the work of Eisner (1985, 1991; see also Eisner & Peshkin, 1990) who argues for both the appreciation of, and aesthetic judgment about, the qualities of programs. While his work has been in education, the ideas are applicable to other sectors. He notes that the role of the inquirer is seeing (more than mere looking at) and appreciating the qualities inherent in programs, "interpreting their significance, and appraising their value [worth]" (1991, p. 1). Drawing on the tradition of criticism, especially in the arts, he takes the position that interpretations and assessments of programs are quite variable and that some are more useful than others. Judging merit, then, should be accomplished by an evaluator knowledgeable about the field – in Eisner's words, a connoisseur.

Appreciating and interpreting the qualities of a program also require specific talents and skills of the evaluator: a skill with language. It is through language, both verbal and non-verbal, that the subtle and nuanced attributes of a program become known and are capable of being shared – communicated – with others. Eisner (1991) argues for voice as well as alliteration, cadence, allusions, and metaphor as part of "the tool kit" of evaluators conducting sustained inquiry whether qualitative or quantitative in method (p. 3). While words are clearly the primary mode of communication, we argue in the final section that there are other equally powerful modes for communicating quality or qualities.

While we are not methodological "purists" (Rossman & Wilson, 1994) or ideologues, we argue that qualitative inquiry may well be best suited for the "seeing'," appreciating and interpreting, and appraising of programs proposed by Eisner. Given its assumptive roots, qualitative inquiry is intended to provide detailed, complex description and interpretation of social programs. Underscoring this argument is the position of Schwandt (1997) who notes that "'qualitative' denotes of or relating to quality, and a quality, in turn is an inherent or phenomenal property or essential characteristic of some thing (object or experience)" (p. 130). He goes on to assert that "Elliot Eisner's . . . explication of qualitative inquiry begins from the point of view that inquiry is a matter of the perception of qualities and an appraisal of their value" (1997, p. 130). When defining qualitative evaluation, however, Schwandt focuses only on the first and second meanings of the term *quality* discussed above: Qualitative evaluation is "a broad designation for a variety of approaches to evaluating (*i.e. determining the merit or worth of, or both*) social and educational programs, policies, projects, and technologies" (1997, p. 130, emphasis added). In an even more reductionist definition, Scriven (1991) asserts that qualitative evaluation is "the part of evaluation that can't be usefully reduced to quantitative procedures" (p. 293). But he goes on to say in a somewhat contradictory manner, "A substantial part of good evaluation (of personnel and products as well as programs) is wholly or chiefly qualitative, meaning that description and interpretation make up all or most of it" (1991, p. 293).

Given the complexities of the term quality explicated above, we argue below that the appreciation and interpretation of social programs and interventions is best accomplished through critical analysis of thick description and sustained reflection on that analysis. Without thick description and reflection on that description, judgments of merit or worth are impoverished. To provide the description and facilitate the analysis – and thus promote utilization (Patton, 1997), we suggest a specific evaluator role: that of *critical friend.*

CRITICAL FRIEND AS FACILITATOR OF DIALOGUE

The role of the evaluator as critical friend offers new possibilities for sustained reflection, critical analysis, organizational learning, and evaluation utilization. The corporate world has recognized this role for many years in its elaborate discussions of learning organizations; leaders of these organizations succeed because they use systematic inquiry to build flexible and trustworthy mental representations (mental models) that allow them to judge merit and worth and, thus, to guide their decision making (see Senge, 1990). As critical friends, evaluators can help program leaders consider the quality (both merit and worth) of their programs by detailing and analyzing the qualities (characteristics and attributes) of the program. This section describes the role of the evaluator as critical friend focusing on the relationship between evaluator and program personnel.

The relationship encourages dialogue, discovery, analysis for change, and small-scale experimentation. It is a critical partnership in which *critical* implies critique, not criticism. Critique offers both positive and negative feedback with the intent of improving the whole. The partners explore *critical*, meaning **essential**, questions, that is, those that explore the heart of the issue and recognize the tentative and speculative nature of any answer. As well, critical inquiry aims to explore **alternative perspectives**. Critical questions are grounded in a social justice framework and seek to discover a more just system. A typical social justice question asks: "Whose interests are or are not being served by this perspective?" Critical inquiry also encourages forward **change**. Derived from nuclear physics, as in "critical mass", and similar to Piaget's notion of the organic process of disequilibration, critical inquiry results in the reorganization of categories and thus, in emergent, new meanings. Partners move from a false equilibrium through disequilibrium to a more grounded state.

The critical friend partnership establishes the conditions necessary for inquiry and evaluation: purpose, trust, a risk-encouraging environment, and willingness to question operating assumptions and existing mental models. The critical friendship coalesces around the purpose, that is, evaluating the program. The evaluation will inform decision-making about the program or about policies related to the program. Trust comes from the mutual benefit derived from collaborating to produce results that can be used. Risk taking is encouraged because the partnership aims to describe and analyze, not to pass final judgment on the program or its leadership. The inquiry is reciprocal in that both parties make their thinking explicit so that it is subject to mutual examination. The evaluator does share judgments, but they are integral to the emergent dialogue, not conclusive statements. Finally, by revealing and questioning assumptions that underlie practices, their dialogue seeks new voices and alternative perspectives. The evaluator as critical friend encourages a willingness to listen to and consider these views. An evaluation can present an ideal site for a critical friendship.

The critical friendship between the evaluator and the program leader is a *heuristic*, or discovery, process. The partners discover or generate knowledge in the form of the program's quality and qualities, and it is the evaluator's role to facilitate this discovery. The tool for discovery is dialogue, a method of communication that uses the stream of meaning that flows between the participants to build shared mental models (see Bohm, 1990). Dialogue surfaces assumptions for mutual consideration and understanding. Dialogue includes explicit communication as well as a deeper tacit communication. Dialogue is reciprocal; the partners make their thinking explicit and subject to mutual examination, thus producing a shared consciousness rather than a conclusion or judgment. In evaluation, the dialogic process results in knowledge that is an *understanding* or interpretation based on the explicit description and analysis of the program.

Traditional models of inquiry rely on **propositional** or conceptual knowledge – intellectual assertions, grounded in logic and organized according to rules of evidence, *that* something is true. Critical inquiry recognizes the multidimensional nature of knowledge, that propositional knowledge is systemically interdependent on other forms including **experiential and practical** (see Heron, 1996). Experiential knowing is "direct, lived being-in-the-world" knowing (p. 33). Practical knowledge is knowing *how*, that is, the exercising of a skill. The various forms of knowledge support, and are supported by, each other. All these forms can be expressed through words or presentational formats using images and patterns such as in plastic, moving, or verbal art forms. The critical inquiry dialogue draws attention to and describes the *experience*, allows its expression or presentation in words, image or pattern, and establishes it as a *proposition* that can shape *practice*. The process results in a deeper and richer knowing *how* to do something grounded in a deeper and richer, more *qualitative*, knowing *that* something is true.

This movement between thinking and doing, and back again, is important in program evaluation if evaluation is to improve practice – to be fully utilized. The critical inquiry process uses this technique. The dialogue permits the surfacing of assumptions drawn from experience. Assumptions, like propositions, are assertions that something is true, but unlike propositions, they are not necessarily grounded in logic nor are they organized around rules of evidence. The dialogue elaborates these assumptions and provides the reflection and analysis that transforms the assumptions into testable propositions that can be prioritized and acted upon.

This dialogic process requires facilitation to ensure the flow of meaning and coherent understanding that generates knowledge. As critical friend, the evaluator takes on the facilitator role. Her inquiry skills are key to building description and guiding analysis. First, her questions and feedback guide the dialogue to name the problem, to focus the inquiry. For example, in a study of a program to identify more accurately students with learning disabilities, the evaluator helped clarify that the question was not "Have we identified too many students?", but "Have we identified the disability accurately so that we may serve the children according to their needs?" The program's merit is to be judged on its goodness in identification. Its worth is its value in serving the needs of children.

Once everyone is clear about what the problem or question is, the dialogue can move to corroborating, elaborating, and gauging expectations. During this phase of inquiry, the evaluator facilitates by listening, asking for clarification, probing, confirming hunches. She encourages the development of perceptions and patterns and the construction of complex images: Who is being identified? What are their disabilities? What are their needs? She enables the program personnel to bracket their perceptions, that is, to hold "in abeyance the classifications and constructs we impose on our perceiving, so that we can be sure to be more open to its inherent primary, imaginal meaning" (Heron, 1996, p. 58). She facilitates reframing, the trying out of alternatives: Are these children learning disabled or are they simply in need of services that can be provided only if they are identified as disabled? Through dialogue, the assumptions are grounded and transformed into propositions.

Then the dialogue moves to exploring action. The propositions are examined – and confirmed or negated – through the detail of thick description. Judgments are made, and propositions are prioritized for action. Through thick description, the evaluator communicates special features and merit, and re-acquaints the program leaders with the uniqueness and complexity of the social program or intervention. Program leaders see the program in new ways. Again, the evaluator facilitates interpretation though questions and feedback, she asks: What are the merits of the program? What is its value? What does this mean for practice? New mental models lead to inspiration, re-conceptualization, and initiation of action.

The dialogue relies on the detailed descriptions or attributes of the program under consideration. The evaluator's role is to ensure that these thick descriptions and alternate descriptions are available and used. Dialogue on the qualities of the program leads to dialogue about quality, both merit and worth, of the program. Thick description and reflection on those descriptions communicates qualities in subtly nuanced ways that permit more finely grained judgments of quality. Through the design and use of a variety of modes of communication, with both the program leaders and the stakeholders, the evaluator as critical friend articulates qualities and quality that shape action. These modes of communicating quality emerge from the multiple ways of knowing that are a part of the critical inquiry dialogue.

WAYS TO COMMUNICATE QUALITY AND QUALITIES

In this third section, we argue that complex social programs demand complex forms of representation by evaluators to both program personnel and external audiences. Quality and qualities are revealed through practice, experience, presentation, and proposition. We offer three modes of communication or forms of representation that hold particular promise for interpreting and communicating the subtleties of quality to a variety of audiences. First, we discuss dialogic representations which include both verbal and non-verbal interactions or conversations. This form of communication expresses experiential and practical ways of knowing. The purpose is to capture the emergent meanings and understandings of participants, managers, or others in the social program.

Second, we discuss narrative or textual representations as well-established, and alternative, modes for communicating both merit and attributes – quality and qualities – of what is being evaluated. Narrative-textual representations capture and express, either conceptually or metaphorically, the propositional, presentational, and at times, practical ways of knowing. These representations include, among others: stories, a day in the life of a program participant, a series of vignettes depicting program interactions, and a series of portraits of significant program features. All have in common the thick description characteristic of narrative representations. Embedded in this description, moreover, are judgments of merit or worth.

Third, we discuss visual-symbolic representations (e.g. video, film, photographs, artifacts, art forms, and demonstrations) as potentially valuable modes for communicating with stakeholders as well as program participants themselves. Through its use of image and metaphor, this mode of representation expresses propositional, presentational, and practical ways of knowing. Visual-symbolic representations may be especially valuable when the program purposes and activities are particularly complex and ambiguous and when participants or stakeholders are marginally literate.

Dialogic Representation

Dialogic representation communicates experiential and practical knowing. As we have described above, the dialogue is part of the evaluation process; it provides an internal way of communicating what is known. Our use of the term dialogue in this context implies more than verbal interactions; dialogic conversations include non-verbal interactions as well. Thus, dialogic forms of communication capture and express the deep tacit understandings that emerge from being there – living – in a setting, from *experiencing and doing*. The evaluator gains an inexplicable sense of the attributes and merit of what is happening in the program. These are meaningfully expressed through dialogic representation that includes the paralinguistic. The evaluator knows more than she can say (see Polanyi, 1962), and much of what she knows derives from non-verbal cues (see Rossman & Rallis, 1998, for a discussion).

The evaluator shares this tacit knowledge with the program people by facilitating the dialogic conversations we described earlier during which concepts are suggested or hinted at, tried out, negotiated. Sometimes the ideas are conveyed through words but just as often through body language and innuendo. We remember a time when Rossman was evaluating a federally-funded program for which Rallis had just become the director. Both of us were present at a training workshop the program offered to teachers; both of us independently were judging the session to be ineffective. We sensed that skills and competencies were not being acquired, but we had no words yet to articulate our emerging judgment, especially since the rhetoric of the training touted its own effectiveness. A glance, a gesture, then a question – from one or the other of us – opened the dialogue that allowed us to define and describe the program's dysfunction – and lead to its eventual correction and improvement.

Dialogic communication is often essential in communicating practical knowledge because skills are not reducible to verbal description. To judge the performance of a skill, the evaluator must either witness it or experience it to capture its quality. Heron (1996) sees the quality of skill as the "knack" (p. 111), the inner key to effective action. The generative aspect of dialogue allows the evaluator as critical friend, once she has witnessed or experienced the skill in practice, to facilitate a conversation that addresses the "knack".

Our process in communicating about the training session was dialogic; we did not know what we knew until we explicated together the complex impressions we experienced. A similar example comes from the evaluation of the inclusion initiative in the urban schools. The special education director's words were inadequate to relate what she knew about how the programs were operating in practice. We had to observe in the schools, to experience the activities and interactions, to witness the exercise of skill. Only then could we converse with her about inclusion's attributes and merits. Our conversations teased out the tacit knowledge into shared understandings of qualities and quality.

Narrative-Textual Representation

A second form of communication, narrative-textual, is useful for representing program qualities and quality both internally to program personnel and to external audiences. These representations, which take the form of either the more traditional report or the more post modern/feminist portrayal, express propositional, presentational, or practical knowledge. The most common format for expressing propositional ways of knowing the attributes and merit of the object of evaluation is the report – formal or informal – that states or asserts specific concepts. These include formative and summative reports, policy briefs, analytic memos and essays, and executive summaries. Versions that effectively convey quality are rich with interpretable description that yields clear and logical judgments about merit or worth.

Propositional knowing may also be expressed in presentational formats, those that portray images and patterns through verbal metaphor. These formats may express the understandings so that they are more accessible to their audiences; their portrayals lift the reader above the routine of daily perception and offer a fresh perspective on events or activities or individuals. Their qualities and quality are revealed through images or patterns inherent in what is portrayed. These presentational texts may also reveal the practical merits of the events or activities if the narratives evoke the impact of the skills and competencies acquired (Heron, 1996). Imaginal narratives certainly portray the "knack" of skill performance better than a traditional report.

Narrative-textual representations in presentational formats include stories, poetry, scripts, portraits, or vignettes. These forms are responses to the stories that are actually lived by participants in the program's operation. The narrative text may be a *reply* or reaction, expressing the emotions observed or felt by participants; the reply may serve as a basis for the critical friend dialogue between evaluator and program personnel or as a finding for display to external audiences. The text may be an *echo*, the evaluator's personal re-telling of the themes she saw or heard. Again, the echo serves as a basis for dialogue or as a display of findings. The text could also be the evaluator's *reflection* that considers and analyzes the stories extant in the program's activities and events (see Heron, 1996, p. 89, for a discussion of story responses). Whatever expression, the text is a metaphor for the images and patterns, and the metaphor may reveal merit and judgments of worth. Audiences often have less difficulty identifying with the metaphorical expression of qualities and quality than with traditional conceptual expressions.

Moreover, these formats, more than traditional reports, explicitly address the question of voice – whose experiences, feelings, skill acquisition, ideas, or opinions are represented? In these alternative narratives, participants in the activities or events may speak directly rather than through an author's neutral or distant voice. Diverse viewpoints and voices may be represented offering more complete descriptions of and multiple perspectives on program attributes. These postmodern/feminist narratives do not privilege a particular dominant voice or interpretation; rather they allow the qualities to surface and determinations of quality emerge from their consideration.

Examples of the use of presentational text abound in anthropology and sociology. Margery Wolf (1992), while she identifies herself as an anthropologist not an evaluator, illustrates the effects of the different forms of narrative representation in her book *Thrice Told Tale*. She offers a fictional account, her field notes, and an analytic essay to represent the same set of incidents that occurred concerning a woman in a Taiwanese village. Each representation, with its different emphasis and detail, communicates a different perspective on life in the village. Were she conducting an evaluation, the judgments of merit embedded in these tales would be made explicit. Similarly, Sara Lawrence Lightfoot's (1983) portraits of high schools in *The Good High School* depict the qualities of each school with bold and accessible images that are evocative to readers.

As evaluators, we often find the use of presentational narratives helpful to our program partners and their audiences. For example, we generate vignettes or short scripts to share our impressions with program personnel. These vignettes and scripts vividly depict the attributes or qualities that program participants experience so that we can consider their merit or goodness in our dialogue and make judgments on them. To express the merits of program activities and events to external audiences, we share "days in the life" of participants. For example, in an evaluation of a girl's education initiative in northern India, we present Sunita, an eight-year-old Indian girl, to illustrate how a teacher-training program can enhance her schooling experience. In sum, narratives, whether formal or presentational text, can powerfully communicate both attributes and merit of a program and provide a basis for judgment of worth.

Visual-Symbolic Representation

Finally, the qualities and quality of the evaluand can be communicated, to both internal and external audiences, through visual-symbolic representations. These representations include: the graphic and plastic art forms such as drawing and painting, sculpture, and dance; theater arts; artifacts; demonstrations and exhibitions. We also include here multi-media and interactive presentations. These may well encompass the graphic and plastic arts, theater, and exhibitions, but they are new media. These new forms typically express experiential or practical ways of knowing, but may express propositional knowledge as well.

As with the narrative-textual representations, the visual-symbolic formats make propositional concepts more accessible to their audiences through metaphor; they lift the reader above the routine of daily perception and offer new interpretations of events or activities or individuals. Attributes and merit are inherent in the images or patterns portrayed. These presentational formats easily reveal the practical merits because they can demonstrate the acquisition of skills; *seeing* a skill performed expresses "knack" (Heron, 1996) directly.

Because they rely on metaphor, the visual-symbolic modes of communication open multiple possibilities for knowing: viewing or participating offers a substitute for experience; description, while non-verbal, is thick and rich; interpretations are emergent and continuous, not conclusive. The representations serve as symbols that can "key" the viewer or participant into a new and different understanding of an otherwise conventional object, activity, or event. For example, a daily activity such as making and drinking coffee together in the morning at the office when seen as a ritual or interpreted as a dance suggests the importance of roles, relationships, and emotions that reveal attributes and merits inherent in the program. The representations are symbols that either sum up or elaborate meaning in a culture (see Ortner, 1973, for a discussion of symbols). Summarizing symbols, such as logos or flags, capture the values and beliefs, the purpose, the commitment of the organization's or program's culture. Elaborating symbols, such as myths, rituals, and ceremonies, sort out complex and ambiguous feelings, ideas, and activities so that they may be understood and shared by many. The visual-symbolic representations used in this mode of communication are themselves insights into the program and its activities and people, insights that the evaluator and program partners can explore for indicators of quality.

Evaluators commonly use visual-symbolic representations to communicate qualities and quality when they draw attention to program artifacts or live demonstrations for program personnel or external stakeholders. Artifacts provide a basis for propositions or assertions about the program that reveal attributes and merit. When we evaluated the inclusion initiative, we drew attention to the displays at the fair celebrating the partnerships between schools and institutions of higher education. Both program director and external stakeholders judged that the artifacts indicated the superficiality of the partnerships. These determinations ultimately contributed to the school committee's judgment of the program's merit and worth when they made budget decisions.

Similarly, evaluators use live or taped demonstrations as a basis for propositional as well as practical ways of knowing a program if one product of the program is skill development. Evaluators of a workshop on math instruction observed video tapes of the workshop participants demonstrating how they applied newly-learned techniques in their classrooms. Through observing the tapes, the evaluators and program director identified and analyzed attributes and merit in instructional practices. A life skills program for women with disabilities taught participants a variety of hygiene and grooming skills, including applying make-up. As the evaluators, we arranged for people from the funding agency to attend a "beauty fair" (which was open to the public) sponsored by the program. Several fair participants demonstrated their skill in putting on make-up. The program's merit was visible in that the women could apply their own make-up as an important life skill. The funders then decided if this skill acquisition was sufficiently worthwhile to fund.

Less common in evaluation, but still effective for communicating qualities and quality, are plays, skits, and dance. Videos dramatizing and summarizing the impact of a program provide the rich description required for interpretation and judgment; they are especially helpful when shown to external audiences to communicate and clarify complex and ambiguous goals and outcomes. Periodically, when working together, we use short skits to crystalize our impressions and to spark a dialogue with program personnel about activities or events that are open to multiple interpretations. For example, we dramatized (for the director and her deputy) several of the rituals practiced in the office of a community arts program. After we all had a good laugh, our critical dialogue surfaced several organizational strengths as well as a weakness that might have been overlooked had the "data" been presented in a more traditional manner. Because they rely less on the written word, visual-symbolic representations call on the more visual and metaphorical ways of knowing; they compel participants and stakeholders to *see* attributes and merit from new perspectives. Thus, they may succeed in communicating quality when language is not possible or adequate.

CONCLUSION

We have argued that evaluators rely on three conceptions of quality, each capturing an essential element of evaluation work. These conceptions describe the attributes of a program which in turn provide a basis for making judgments about the merit or worth of the program for participants, staff, a community, and/or society generally. We have also argued that, to fulfill the demands of all three conceptions, evaluators should consider the role of the critical friend in which dialogue, reflection, and interpretation blur the historic power imbalances in the relationship between evaluator and evaluand.

The role of critical friend offers a variety of ways to uncover, describe, and communicate the various forms of knowledge generated in complex social programs. These forms of knowledge include propositional, experiential, and practical. Traditional evaluations have focused primarily on propositional knowledge to the exclusion of the others. The role of the critical friend encourages expression and validation of the more subtle ways of knowing.

Finally, we have argued that the complexity of knowledge generated through social programs can be well-communicated through alternative modes of representation. Dialogic representation helps the evaluator and program personnel surface tacit understandings and subtle impressions about a program. These can then be examined for their usefulness in making judgments of merit and worth. Narrative-textual representations of a program's attributes and merit offer new ways to communicate with both internal and external audiences. Likewise, visual-symbolic forms encourage engagement with the program's attributes in forms that appeal to both cognitive and affective reasoning.

In this chapter, we have argued for more complex assessments of social programs, ones that mirror and honor the subtleties and ambiguities inherent in such programs. The role of the evaluator as critical friend encourages this complex reasoning and engagement with a program, its staff, and participants over time. It does, however, demand a commitment to dialogue and critique, a willingness to suspend judgment of merit and worth until a program's multiple and subtle attributes are well known, and an openness to sustained collaborative inquiry. Quality is a complex concept; identifying and communicating quality in evaluation requires a process that recognizes and honors its complexity. The evaluator as critical friend can facilitate this process.

REFERENCES

Bohm, D. (1990). *On dialogue*. Ojai, CA: David Bohm Seminars.

Brown, L. (Ed.) (1993). *The New Shorter Oxford English Dictionary*. Oxford: Clarendon Press.

Eisner, E. W. (1991). *The enlightened eye: Qualitative inquiry and the enhancement of educational practice*. New York: Macmillan.

Eisner, E. W. (1985). *The educational imagination* (2nd ed.). New York: Macmillan.

Eisner, E. W., & Peshkin, A. (1990). *Qualitative inquiry in education: The continuing debate*. New York: Teachers College Press.

Geertz, C. (1973). *The interpretation of culture: selected essays*. New York: Basic Books.

Heron, J. (1996). *Co-operative inquiry: Research into the human condition*. London: Sage.

Lightfoot, S. L. (1983). *The good high school: Portraits of character and culture*. New York: Basic Books.

Ortner, S. (1973). On key symbols. *American Anthropologist*, *75*, 1338–1346.

Patton, M. Q. (1997). *Utilization-focused evaluation*, 3rd edition. Thousand Oaks: Sage.

Polanyi, M. (1962). *Personal knowledge*. Chicago: University of Chicago Press.

Rossman, G. B., & Rallis, S. F. (1998). *Learning in the field: An introduction to qualitative research*. Thousand Oaks: Sage.

Rossman, G. B., & Wilson, B. L. (1994). Numbers and words revisited: Being "shamelessly eclectic." *Quality and Quantity: International Journal of Methodology*, *28*, 315–327.

Schwandt, T. A. (1997). *Qualitative inquiry: A dictionary of terms*. Thousand Oaks, CA: Sage.

Scriven, M. (1991). *Evaluation thesaurus* (4th ed.). Newbury Park, CA: Sage.

Senge, P. (1990). *The Fifth Discipline*. New York: Doubleday.

Wolf, M. (1992). *A thrice told tale*. Stanford, CA: Stanford University Press.

9. CULTURE, STANDARDS, AND PROGRAM QUALITIES

Saville Kushner

THE ARGUMENT

> "*This is a sharp time, now, a precise time – we live no longer in the dusky afternoon when evil mixed itself with good and befuddled the world*" (Deputy-Governor Danforth in Miller's (1976) *The Crucible*, p. 76).

Precision is a passion. Like all passions it is in its nature to overwhelm. We, too, live in "sharp" times in which the corruption of criterion referencing (Glass, 1978) has led to an overwhelming and unreasonable certainty in the criteria against which educational achievement and action are judged. Society and its representatives too easily believe the myth of discrete standards. The zeitgeist we evaluators inhabit is analogous, in this respect, to the one that led to the ravaging of Salem – a belief in "excellence", in "best practice", in "bench-marking", in objectives accomplishment and in the mere possibility of agreeing a definitive moral standard. We are still burning witches.

We have lost, in Glass's (1978) terms, the procedural interpretation of standards – i.e. standards merely as guides to position and progress – and with it a tolerance for an essential lack of precision in invoking them. "*In education,*" he argued shifting the debate from fixed levels of achievement to a discussion of changing states, "*one can recognize improvement and decay, but one cannot make cogent absolute judgments of good and bad*" (p. 259). However, in the hands of an accountant or an economist an educational "standard" has come to mean a measured target for purposes of justification – the closure of process.

Visions of Quality, Volume 7, pages 121–134.
2001 by Elsevier Science Ltd.
ISBN: 0-7623-0771-4

A standard implies the dismissal of alternative judgments and a requirement to stop thinking about an educational proposition. This has, itself, allowed for a corruption of conceptions of educational quality, which comes increasingly to stand for the reduction to a single unit of assessment of many complex features of a culture – a program, say. Once we have yielded to the notion that we can dismiss alternative measures of educational worth we can simplify the measurement of a culture – often with breathtaking precision. I have just moved from one university research center, nationally Graded 5 for research quality, to another graded 3b – on a scale of 1 to . . . well, does it matter? The fact is that, in resourcing terms, on February 28th 1999 I was a Grade 5 researcher; on March 1st I became a Grade 3b researcher.

Both terms – "standard" and "quality" – thus laundered of their complexity – combine to create bedrock on which summative judgments can be made regardless of context. It is hard, these days, in our technocratic version of a theocratic state to use these terms in alternative ways, as we have been more and more closely scrutinized for our compliance with external targets, for our homogeneity and for our abrogation of complexity. In this chapter I want to look critically at these two terms and to do so I will describe a particular evaluation case. The account that follows is taken from an enquiry that fell within the rubric of "Democratic Evaluation" (MacDonald, 1987). This case study approach to evaluation consists of the design of a moral and political order within which there can be an open contestation over the meaning and aims of a program. House (1980) characterizes it as a "transaction" approach, focusing on the information – exchange element of the model, through which evaluation knowledge is diffused across a program and irrespective of hierarchy. Democratic Evaluation is designed to deny any particular group the privilege of dominating the evaluation agenda, and so to reveal and legitimate plurality.[1] See the rather long footnote for a comparative account of the approach alongside the more familiar empowerment evaluation with which it should not be confused – but in terms of my agenda in this paper hold in mind that Democratic Evaluation is concerned to return the setting of criteria for defining quality and deliberating over standards to the level of experience and action. The democratic evaluator travels light of formal theory and political direction, and seeks her agendas from program participants and stakeholders.

A key aim of this approach is to put evaluation to the service of self-determination, which itself implies promoting the realities of those who "people" programs and asserting the importance of context. Elsewhere, I argue (Kushner, 2000) that the social program has become a common means of imposing extrinsic political realities in people's lives and undermining the capacity for self-determination – and that the business of a socially just evaluation should be to challenge that by inverting the relationship between program and individual. In place of the common practice of documenting a program as a context within which to "read" people's lives, we should, I argue, be documenting people's lives as context within which to examine the significance of a program. The question is how programs and their participants are suffused with each other's meanings – often in tension – and how to strike an appropriate and just balance. At stake are the criteria each would bring to bear on the setting of standards for success and failure.

We will see, when we come to the case, how this can be worked out in practical terms. For now, I am concerned to show the link between a discussion of standards and quality, and the task of creating the kind of ethical space within which individual voices can be heard, and within which a program's ambitions might be judged against the world-views and the meanings of its participants. Obviously, I will be claiming that the use of precision and exclusive definitions in enunciating standards and "quality criteria" is a means of denying self-determination, and of imposing values and meanings derived from political cultures whose interests and values might (too often do) diverge from those who are charged with implementing the standards.

The evaluation I will use to exemplify my argument – of bilingual schooling policy in the USA – was attached to a new moral order in a fairly concrete way. The Ford Foundation, seeking to take a leadership role in debates about bilingual education policy, came to England and to MacDonald's evaluation group seeking an alternative (if not an antidote) to forms of evaluation based on conventional theory-driven approaches that had done little service to the cause of bilingualism – either in treating it with sympathetic judgment or in providing it with practical insight. These approaches to evaluation had widely been seen in the bilingual movement at the time as having appropriated the meaning of the civil rights struggle that underpinned the educational endeavour. MacDonald's group was approached as much as anything because we were prominent advocates of case study evaluation, which was a strategy for legitimating the experience and values of program participants.

In the account that follows I will try to portray evaluation as it moves about in the "*dusky afternoon when evil mix[es] itself with good and befuddle[s] the world*" (Miller, 1976). This is an account from inside a Democratic Evaluation.

THE CASE

I'm sitting in the dark, windowless staff room of the Rafael Hernandez School in Boston having my haircut and I'm thinking that I'm fairly smart. She is a

Puerto Rican teacher-aide who is cutting it and we are getting along pretty well – she's joking loudly and orchestrating the good humour in the room among the other teachers. *Por aquí todo anda bien.* All are heartened by my vulnerability as the educational evaluator, along with his hair, is cut down to size. The school has rejected the first draft of our portrayal of it somewhat dismissively and my job is to renegotiate access so that we might construct a more acceptable account. I seem to have won them over once more – *por lo menos, les gusto.* I'm thinking of how we might get back into the classroom.

In a case study-based evaluation of bilingual schooling policy we had placed this magnet school – reputedly a fine example of a fully bilingual program – at the center of the case. However, the means we had used of comprehending the work and accomplishments of this school were the instruments of sociolinguistics, linguistics and transactional analysis, which had thus far, in their terms, failed to do justice to their efforts. The account was, for the most part, observation-based and we had focused on language interactions. These had, hitherto, been the means employed by the scientific community to evaluate bilingual education and people in this school and others associated with it alleged that the methodology had consistently failed to appreciate the true nature, accomplishments and qualities of the movement. We were denounced and dispatched amid allegations that we had failed to grasp what it was that made the Rafael Hernandez mission and practice so distinct.

We were, in fact, to return to the classroom and to create a new database for another attempt at a portrayal. Throughout the process of renegotiating access, however, we were always in danger of reigniting the hostility of the school. Time and again we were told not to look in the classroom – after all, what would you expect to find there other than the sterile, context-less mandate of the special program and the prescribed curriculum? "*Ven al picnic*", they'd say, "come to the party on Friday – to the prize-giving! *Ahí se lo puede ver!*" We were invited to the sports day, to the playground, told to linger in corridors. *Nos invitaron andar en el mundo informal de la escuela.* But we resisted, protesting that the caring and affection they wanted us to observe – and which we saw – were merely the baseline we would expect from any school in charge of young children and that this hardly qualified as a distinct educational mission – certainly not as a substantial curriculum program. We held out.

But, of course, it was Carmen who was winning me over as she sliced with relish into my *peinado*. I had our relationship the wrong way round – *fue un chantaje – la vida te da sorpresas*. Eventually we came to be persuaded and left the classroom to document the "informal" curriculum – in fact, the only real space left to these innovative educators to discover a self-directed curriculum program; actually, one connected more to the history of that community's struggle than to socio-linguistic theory or Washington policy. Classroom processes were dominated by the dry, regime of entitlement programs, behavioral treatments and special programs – all conceived in professional cultures well beyond the school and all wedded to standards generalized to distant contexts and ambitions. Outside of the classroom, however, the curriculum specification was more appropriate to the circumstances of these children – the stresses caused by multilanguage experiences (usually in a context of poverty) created certain emotional needs in young children: respond to these, discover a respect for children and the rest will flow. Social relations based on mutual respect created a framework for curriculum relations. We only returned to the classroom once we had understood this and could comprehend the school in its own terms. Thereafter, our task was to document the struggle between externally imposed standards and internal convictions about what counted as worthwhile educational interactions.

When we did return to the classroom it was to ask very different kinds of questions than about language pedagogy, and to do more than observe pedagogical interactions against a theoretical framework of understanding. In fact, observations were used as underpinnings for subsequent interviews, questioning teachers about purposes and means – and were conducted on a broader range of lessons. We were able, for example, to reconstruct a theory of bilingual pedagogy from observations and discussions of how teachers taught on a daily basis in other than language settings. Here we sought to document the standards teachers set themselves and their children in taking on the bilingual challenge, and we were able to critically review the theoretical underpinnings of our previous work. We developed such a grounded theory – more accurately a practical theory – which we were able to contextualize in the quotidian life of the school. This theory embodied a critique of extant approaches to theorising about bilingual education. For example, much attention had been paid in the research literature to "stigmatization" of what was commonly known as "phonological interference", borrowing and "non-standard" English. Equally, much research attention had been paid to the dysfunctionality of "code-switching" – i.e. moving between two or more languages in the same utterance.

We subsequently argued on the basis of our experience that conventional approaches to language education and language acquisition were inadequate where they failed to take into account social context – an insight achieving the status, today, of the commonplace. At one point in our analysis we said that:

> *Where social integration is high, where social relationships between teachers and students, and students and students, are ones of mutual respect and even affection, and where parental support for the school is high, then what have been taken as prominent areas and issues*

> *for research into bilingual education are diminished in importance* (MacDonald, Adelman, Kushner & Walker, 1982, p. 220).

Here, implicitly, was a challenge to theoretical standards. We noted that, left to their own devices, teachers would use a range of pedagogical forms – from the instructional to the "culturally responsive" – within a teaching logic that their own values and approach made coherent. We further concluded that:

> *Without understanding the relationship between social interaction processes within classroom and school, and the nature of the curriculum and its organisation, the consequences of legislation or administrative decisions on schools and school systems can neither be predicted nor even monitored* (MacDonald et al., 1982, p. 222).

If the consequences of curriculum policy decisions on schooling cannot be predicted, at what price a curriculum standard or a behavioral objective?

We were, therefore, able to do more than derive practical pedagogical theory. The case study was not of the school, but of bilingual policy. We were taking a "slice" out of the cake of the Federal/State political and educational system, a "slice" that cut across the various levels of action and policy response including at school level. The school lay at the heart of the case but did not define it. This meant that we had to discover contingent relationships between what happened in the school and beyond it – for example, in the Puerto Rican community that created and then lost control of the school and in the political community that fêted and then co-opted bilingual community activists. The range of such connections we could make had previously been restricted by an overly narrow focus on formal curriculum and pedagogical interactions. But this re-focus on practical, embedded meanings meant that we were able, not just to track policy as it was translated "down" the system into practitioner settings – the conventional "direction" of evaluation accounts, but also to start with practical needs and constraints and track these "up" to the point where they came up against the restraint of policy. This evaluation would evaluate standards, rather than hold people to them.

What we needed to do to broaden the range of such connections was to get into values. What we had discovered was, on reflection, unsurprising – that some key teachers in the school simply did not share the values base of the curriculum they had, perforce, to teach. Having spoken to teachers about where their values did lie we were in a position to identify and explore the contradictions and the tensions – to reach beyond what was (defined by formal curriculum standards) to what might have been (defined by internal aspirations). The pedagogical theory we did arrive at through joint analysis with teachers was, as we might have expected, at odds with the curriculum they were expected to teach. We were able to expose a clash over what counted for reasonable standards to hold for the educational endeavours undertaken by the school: we were able to portray the interface between the generalized/generic standards promoted by the formal curriculum, and the particular/situated standards being generated inside the school.

Here was a significant clue as to the importance of what we were told were "extra-curricula" activities – what we saw as non-classroom-based pedagogy. The classroom was, in many senses, a hostile environment for teachers (let alone children). The array of accountability requirements on teachers, the necessity to teach the range of special programs, the frequent incursion of non-school professionals empowered to intervene and monitor the children, the emphasis on behavioral objectives – all of these had the twin effects of: (a) subverting the teachers' judgments about what counted as a wholesome learning environment and an educational standard, and (b) fragmented what otherwise might have been a continuous and coherent curriculum experience. The school had made an ingenious and courageous attempt to regain control over a hopelessly diverse curriculum with what they termed a "levelling system" – i.e. designing individual pathways through the special programs, often cutting across year-groups. But this was not sufficient to customize the formal curriculum, and teachers were forced to retreat from the classroom to reassert what, to them, were meaningful educational interactions.

Hence, at school sports days and school picnics the teachers could interact with children and parents, lose races with children, engage freely in banter with children, wear less formal clothing, be challenged socially by children backed up by their parents. These were experimental settings, but settings in which children could substitute compliance with teacher authority with mutual caring and respect. Here was the rich and culturally coherent learning environment the teachers were after but were denied in the classroom. When they urged us to look beyond the classroom they were asking us to observe their relatively autonomous attempts to freely manifest their curriculum values – i.e. to explore what, for them, constituted "quality" in their educational endeavours. *Pura vida.*

What Was Happening Methodologically?

There are a variety of ways of characterising the methodological shift we undertook to complete this more pluralistic portrayal of the school. Certainly we moved from a somewhat theory-driven methodology to a responsive one; and, also, we were moving away from a discrete curriculum evaluation strategy back to the policy evaluation we intended all along – i.e. we were able to use the situated discourse of teachers to hold policy to account, as well as the more conventional approach of supporting judgments of how faithfully practitioners realized policy ambitions. At the center of it, however, was a shift of focus from operations to culture – from performance to logic. The principal methodological shift was away from externally driven criteria of judgment to internally determined criteria. Both were theory-driven – but we moved from a reliance on formal theory to "practical" theory derived from teachers' own analyses.

Nor was this merely a matter of taking more account of context – in fact, it made the very notion of context problematic. We always had the view that whatever we observed in the school had to be located within some sense of the surrounding and supporting community. Rather, the view the school persuaded us to have was one in which context and focus – action against its contingent background – were connected by a blurred boundary if they were not, in fact, seamlessly continuous. As we argued in our concluding analyses, we did not merely make the archetypal move from the formal curriculum to the "hidden" curriculum. We were, in fact, taking a more complex view of curriculum as an integrated experience. Our new portrayal included biographical detail on teachers, social and political relationships, out-of-school activities, and apparently unrelated curriculum activities.

The schoolteachers were insistent that there was a strict division between what happened in the classroom and what happened outside it. The former represented the meanings forced on them by designers of "official knowledge" (Apple, 1993); the latter represented their own "unofficial" meanings. We respected that and called the former the "formal" curriculum and the latter the "informal", but more lay beyond the distinction. The key point of contestation in the community of bilingual advocates lay between "transitional" bilingual education and "cultural maintenance" education. The former denotes the use of bilingual programs to ease the entry of non-English mother-tongue pupils into the mainstream curriculum and into, therefore, the competition for employment opportunities. The latter – "cultural maintenance" education – saw bilingual programs as cementing and enhancing the cultural and linguistic background of those children in a context where every other social influence threatened it. Success for the former might be found in better achievement scores in mainstream education and better employment rates among (in this case) Hispanic children; success for the latter advocates might be represented by more, and more meaningful conversations between children and their Latin grandparents. In retrospect, there was a more fundamental difference. Transitional aspirations focused on attainment, on the mastery of minimum competencies on which might be based a bid to join the mainstream vocational culture. Maintenance, however, relied more on conversation and on the vagaries of understanding. Each has a very different approach to standards and quality. Transitional bilingual education relies heavily upon minimum competencies and, therefore, upon the (corrupt form of) criterion assessment; for maintenance education such forms of assessment are meaningless – how do you measure, against a common criterion, the capacity for children to engage meaningfully with their grandparents?

In the end, we argued the need "*to reunite Language Arts and school picnics*" – not merely as "connoisseurs" of curriculum, but on the basis of systematic observation that this integration was a task that faced the child every day at that school, and that it merely made sense for teachers to do the same. Some of the consequences of bilingual education policy were that teacher-judgment was subverted and curriculum exposed to the values and logics of unseen and unknown persons, that curriculum was fragmented, that the classroom was not designed to embrace multiple purposes. The result was that pupils were left to discover a sense of coherence – to integrate the many experiences as they moved in and out of the formal and informal curricula and in and out of special program classes.

So another aspect of the methodological shift was away from an allegiance to the vocabularies of articulate power groupings (language theorists, for example) and towards advocacy for the marginalized voices of children and teachers. This is not to say that we sought to discriminate against one group, but that we sought to discriminate in favour of another. This echoes House's (1980) view that justice lies in "fairness" and that the latter justifies positive discrimination. What it allowed, of course, was a broader range of possible perspectives on the case and a respect for the right of individuals to choose their own meanings – at least in the context of the temporary moral order created by the evaluation. But the findings of the evaluation proved difficult to incorporate into the bilingual policy discourse of the time – not simply because President Reagan eliminated bilingual program funding, but perhaps also because it challenged the very notion of standard-setting – and, therefore, systems of political control.

What Was Happening in Terms of "Quality"?

There is another way in which to understand this shift and it is this to which I want to refer now. I have said that we moved away from a theory-driven evaluation. The very use of a methodology that measures practice against theory is searching for some sort of measured index of "quality" – i.e. there are at least implicit standards to be reached and these are imported by the methodology as it seeks to arrive at a calculus balancing the richness of experience against the requirement to meet expected measures of performance and impact. The more obvious version of this, of course, lay in the classroom itself where other standards (also derived from theory) were being imported in different ways – largely through behavioral objectives-based curricula. In our early portrayal we had a similar status in the eyes of teachers as other (white) science-based authorities seeking to control educational practice rather than understand it.

The move to observing the informal curriculum – i.e. practices informed by the internal standards of the school – was a shift away from a concern with the "quality" of the program and towards its "qualities" – i.e. in the more common-sense use of the term denoting nothing more or less than what makes something distinctive. Yet another way, then, of characterising the methodological shift we made was from using "quality" as an explanatory construct (what made this a good program) to using quality as a descriptive construct (what makes this program distinctive). It is this shift I want to talk about, and I will do so in relation to cultural theory. Here, I will treat a program – in this case, the bilingual education program as represented by the policy and the practice – as a culture.

Stenhouse (1967) viewed culture as "*the medium through which individual human minds interact with each other in communication*" (p. 13). His preoccupation as an educator and curriculum theorist was to understand the cultural basis of curriculum and pedagogy in light of the central tension between the individual and the group and how people shared and differed in their allegiance to certain beliefs and practices. A central part of Stenhouse's understanding was the concept of the "standard". By standard he referred to "*criteria which lie behind consistent patterns of judgment of the quality and value of the work*" (p. 70) – i.e. a standard was a procedural principle. The "medium" for communication was regulated, and regulated in such a way as to produce consistency in judgments about how to proceed. Elsewhere, Stenhouse (1963) talked of this "consistency" in communicative interaction as "*a climate of expectation . . . all action takes place within this climate*" (p. 124). It is that notion of consistency – the possibility of expectation – which evaluators seek when they look for the qualities of a program and which we are always in danger of interrupting by introducing our own (or other's) view of what is significant or relevant. This is what we threatened to do in the Hernandez School with our first draft, and what we veered away from when we moved to document those practitioner-level "agreements" over what was worthy of discussion and action. Outside the classroom was a set of cultural activities through which the school could explore its distinctive contribution, try out novel relationships, interact in communication. Communication (between professionals, children and community) was fragmented by the formal curriculum.

The danger, as in the case of the Hernandez School, is when evaluators import a view of consistency that owes more to cultural standards beyond the case than to those within it. Anxiety at working in a research environment which at first often seems undifferentiated and, therefore, disordered often persuades us to rush, not so much to judgment, as to an assertion of consistency, which in truth may be more of a comforter for the evaluator than a reflection of real meaning. In fact, I want to substitute "coherence" for "consistency", for I think this is closer to what Stenhouse was aiming at. Consistency implies a uniformity and a rationality to the order of things measured against an external or meta-criterion; whereas coherence allows for events, for example, being rational as measured merely against each other in their context (Everitt & Fisher, 1995). The shift we were forced to make by the Hernandez staff was a shift from our theory of coherence to theirs; from a platform for judgment created by people far from the context to one created by those who lived and worked in the context.

Here, then, we invoke an alternative voicing of the concept of a standard – one which, as I have suggested, relates less to the quality of a program and more to its *qualities* – i.e. less to its success and failure against a criterion and more to the range of its social and political facets. We could talk about the quality of a classroom, for example, anticipating a judgment as to its being a "good" or a "poor" educational setting against some criterion that classrooms share in common with others. When we talk of the qualities of a classroom, however, we merely seek to characterize it – to make it distinctive from others. Let me quote Stenhouse (1967) on the matter:

> *When we say that we regard the work of one class as better than that of another, we are not simply judging it to reach a 'higher' standard. Such a conception implies a common measure against which both classes can appropriately be assessed, but in fact standards can be qualitatively different. When they are, a comparative assessment is not a matter of measurement but a matter of value-judgment. For example, we may opt for creativity or correctness. . . . Such choices are founded upon conviction rather than demonstration. The sources of standards in school work lie in the teacher's values* (p. 75).

The standards he was referring to, of course, were the standards that have been derived from inside the school, but also inside each set of pedagogical interactions as a source for those criteria that regulate the communication medium – i.e. which define each classroom or picnic as a culture (or sub-culture). So the "qualities" of the bilingual program we were looking at had less to do with pedagogical compliance with theoretical principles and more to do with the diverse expression of teacher values in each classroom and non-classroom setting. Our job was to document that diversity and to understand the source of convergence. The possibility of an overall, summative measurement of "quality" of the program (the school curriculum) was, at best, an irrelevance and at worst a source of distortion.

We can see, now, that what the evaluation was guided into by the school was a methodological journey, which took us through the values territory of the school and its practitioner-theorists. The school had, mostly implicitly, a theory of itself – i.e. it was a self-conscious culture in that it had standards that represented a belief system and against which it could measure the coherence of the actions of its members – and, too, the coherence of their own aspirations against those of the political groups massing outside the school boundaries. Interestingly, we were challenged once again over our use of the word (and the concept) of "politics" within the school. There was what to us was a clear case for a political analysis of the school (all fully-paid teachers and the Principal were white or Cuban; lesser-paid teacher-aides were from poorer, generally Central American countries, for example), but there was open hostility to our using the term. The school was anxious not to allow another conduit through which their world of meaning might be colonized once again by the political communities beyond. We yielded.

SUMMARY

The simplest, unfortunately least effective, "defense" against Deputy Governor Danforth is, of course, to say "well, I just don't see things like that". In Salem there was no mediating device between the wielders of judgment and those whose lives were to be judged, for the judges were putative agents of a higher, immutable logic. Hence, there was no appeal against the arbitrariness of the theocratic judgment – "*As God have not empowered me like Joshua to stop this sun from rising, so I cannot withhold from them the perfection of their punishment*" (Miller, 1976, p. 104), bemoaned the hapless Danforth. The standard is the standard. This, in fact – the absence of mediating structures – partially defines a theocratic state, a state that asserts a unitary system of meaning. But the liberal democratic state is partially defined by precisely its opposite – the presence of such mediating structures and the possibility of some distance between those who judge and those who are judged. An element of a liberal state is the live practice of holding logic of government publicly accountable; the increasing absence of this condition in Western democratic states is corrosive of liberalism.

Evaluators inhabit that mediation territory – which is why it is so important for us to maintain some distance from each side of the judgment-equation (and, hence, Democratic Evaluation – in its best expression – refuses to make recommendations). Contemporary Western political cultures are fearful of such arbitrary power and vest some vestigial democratic authority in intermediary agencies. Evaluation – to greater and lesser degrees politically neutral – has thrived on that. Perhaps, as our democracies grow more liberalized and less liberal that ethical space in which we conduct our business becomes more straitened – perhaps, too, we are, as a community, increasingly co-opted into political ambitions. Nonetheless, there is at least an opportunity at the margins to play a role in enhancing the self-determination of people and increasing the accountability of policies and programs to that aim. Central to the task of such an intermediary role is resistance to the passion of precision. There are many ways to define standards and there are many approaches to understanding quality, and it is the responsibility of the impartial evaluator to resist the arbitrary dismissal of alternatives. We should treat with professional scepticism the increasingly common claims that such concepts as "national standards", "best practices", "quality control criteria", "benchmarks" and "excellence in performance" are meaningful in the context of professional and social action.

Glass (1978) warned evaluators and assessors against ascribing a level of precision in judgment to a subject matter that has less of it in itself – here evaluation contaminates its subject. Programs are rarely as exact in their aspirations, processes, impacts or meanings as our characterisations of them and as our measurements of their success and failure. Glass urged evaluation to avoid absolute statements and to stay, at best, with comparative statements – rough indications of movement, what makes one state or position distinct from another, distinguishing ascendance from decline, etc. Given the 20-odd staff at the Rafael Hernandez School, the 200-or-so pupils, the range of languages and social backgrounds, the plurality of meanings perceived in a curriculum statement – given all of these where is the boundary between the exact and the arbitrary? And if we are to settle for the arbitrary, why commission someone as expensive and potentially explosive as an evaluator? Ask a cleric.

NOTE

1. To counterpose it with "Empowerment Evaluation" (e.g. Fetterman, 1997), with which it seems increasingly to be confused, the model of Democratic Evaluation proposed by MacDonald is more an approach to disempowerment in that it seeks to neutralize the distorting effect of political status in a program – i.e. there is an assumption of a zero-sum in relation to power – that the "empowerment" of some presupposes the "disempowerment" of others. Where Fetterman is concerned with the empowerment side of the equation (essentially using action-research approaches in an iterative process of action and reflection), MacDonald's Democratic Evaluation tends to be more preoccupied with the creation of the ethical and political space within which we might engineer some momentary disempowerment – "Empowerment Evaluation" is relevant to a process of

institutional self-evaluation where Democratic Evaluation is best achieved through external evaluation. Where Fetterman defines democracy in evaluation as "participation", MacDonald defines it as much in terms of restraint, as it were, on "over-participation" by power elites (not to dismiss the rights of the powerful in evaluation, but to equalize them with those of the less powerful). In that sense, Democratic Evaluation shares with Empowerment Evaluation a concern to reflect our contemporary moral order which, to quote Berger (1974) "implies the right of the individual to choose his own meanings" (p. 195) – in that one task of empowerment and democracy is to remove constraints to personal meaning and to subvert the process of imposing extrinsic meanings on people's lives. For Fetterman, empowerment is about the self-determination of social groups; MacDonald has been at least as preoccupied with the task of preventing the self-determination of power elites to displace that of the relatively less powerful – i.e. concerned at least as much with "between-group" as "within-group" realities. But both MacDonald and Fetterman share a concern to reduce to levels of direct experience and professional action the responsibility for setting criteria for what counts as wholesome and worthwhile action; both, too, are concerned to focus judgment of quality and discussion of standards on process at least as much as on outcome; and both assert the necessity for evaluation to respect the integrity of context – i.e. that the meanings which count are those which are held and determined within the embrace of direct experience.

REFERENCES

Apple, M. (1993). *Official knowledge: Democratic education in a conservative age*. London: Routledge.

Berger, P. (1974). *Pyramids of sacrifice: political ethics and social change*. London: Allen Lane.

Everitt, N., & Fisher, A. (1995). *Modern epistemology: A new introduction*. London: McGraw-Hill.

Fetterman, D. (1997). Empowerment evaluation and accreditation in higher Education. In: E. Chelimsky & W. R. Shadish (Eds), *Evaluation for the 21st Century: A Handbook* (pp. 381–395). London: Sage.

Glass, G. (1978). Standards and criteria. *Journal of Educational Measurement*, *15*(4), 237–261.

House, E. R. (1980). *Evaluating with validity*. London: Sage.

Miller, A. (1976). The crucible: A play in four acts. Penguin.

Kushner, S. (2000). *Personalising evaluation*. London: Sage.

MacDonald, B., Adelman, C., Kushner, S., & Walker, R. (1982). *Bread and dreams: A case study of bilingual schooling in the USA*. CARE Occasional Publications No. 12, CARE, University of East Anglia, Norwich, U.K.

MacDonald, B. (1987). Evaluation and the Control of Education. In: H. R. Murphy & H. Torrance (Eds), *Evaluating Education: Issues and Methods* (pp. 36–48). London: Paul Chapman.

Stenhouse, L. (1963). A cultural approach to the sociology of the curriculum. *Pedagogisk Forskning* (Scandinavian Journal of Educational Research), 120–134.

Stenhouse, L. (1967). *Culture and education*. London: Nelson.

10. DETERMINING PROGRAM QUALITY OF A SOCIAL SERVICE COLLABORATIVE USING CASE STUDY METHODS AND MOVING TOWARD PARTICIPATORY EVALUATION

Hersh C. Waxman, W. Robert Houston and Lynn Cortina

INTRODUCTION

Participatory evaluation is an evaluation method that has been used in many evaluation projects (Cousins & Earl, 1992, 1995a, b; Upshur & Barreto-Cortez, 1995; Whitmore, 1998; Whyte, 1989, 1991). Because it is grounded in the experience of staff, clients, and participants, participatory evaluation is more likely to provide information that is useful to program administrators and policy makers. Participatory evaluation is an approach that is more appropriate for a formative, rather than a summative evaluation, and is more flexible than traditional evaluation approaches. Participatory evaluation often results in positive changes within an organization or project, including increased communication between staff members, beneficial effects on program development, and higher

Visions of Quality, Volume 7, pages 135–151.
2001 by Elsevier Science Ltd.
ISBN: 0-7623-0771-4

quality evaluations (Upshur & Barreto-Cortez, 1995). The primary benefit of participatory evaluation is that it enhances the ownership, relevance, and usefulness of the evaluation for its intended users (Brandon, 1998; Cousins & Whitmore, 1998).

One of the major disadvantages of participatory evaluation is that it is very difficult to get all the various stakeholders to participate in the development, implementation, and interpretation of the evaluation because of the great time commitment that is needed. Consequently, participatory evaluations are often poorly implemented and not truly participatory because the collaboration, input, and commitment from all parties involved in the evaluation is lacking. The initial evaluation reported in this chapter can similarly be criticized because the stakeholders have not been actively involved in all aspects of the evaluation. Yet, although typical case study methods (Merriam, 1998; Stake, 1995; Worthen, Sanders & Fitzpatrick, 1997) have been used for the initial phases of our evaluation, the potential and promise of a "true" participatory evaluation still exists. It is the potential for movement in this direction and our commitment to it that guides the present chapter as well as the actual evaluation.

This chapter describes the use of case study methods and participatory evaluation in evaluating a collaborative project, The BRIDGE/EL PUENTE (*Building Resources Internally for Development and Growth for Everyone/ Pueblos Unidos En Tareas Educativas*), which is part of the Greater Houston Collaborative for Children (GHCC). The GHCC was created to maintain a structure by which the Greater Houston community could develop collaborative thinking and initiatives about children and families that affect the common good. The two major goals of the GHCC are: (a) to promote creativity within and among organizations and empower the community to find new ways of solving old problems, and (b) to pave the way for a more effective child-serving system in terms of both policy and practice. The first major undertaking of the Collaborative was to launch a funding initiative that focuses on children from birth to age six and their families, and requires collaboration on the part of interested service providers. One of the collaborative projects that was initially funded is The BRIDGE/EL PUENTE project, which is led by the Neighborhood Centers Inc. and the Gulfton Community Coalition.

In this chapter, we first describe The BRIDGE/EL PUENTE and then the case study methods and participatory evaluation approach we have used for the initial phases of the project. Next, we examine perspectives of program quality from several different constituents in the project, such as funders, project staff, service providers, and the families being served in the project. Finally, we report on the advantages and disadvantages of determining program quality through the participatory evaluation process and we highlight some of the evaluation lessons we learned.

DESCRIPTION OF THE BRIDGE/EL PUENTE PROJECT

The BRIDGE/EL PUENTE project is situated in a large apartment complex, Napoleon Square, which is located in the Gulfton area of Houston, Texas. The project was designed to help parents with children from birth to six years of age who live in Napoleon Square by targeting a variety of services to them through collaboration among various agencies. In other words, the project was designed to bridge the gap between living in poverty and becoming self-sufficient.

The Gulfton Area is home for many immigrant families who have come to the United States in search of a better life for themselves, but most especially for their children. Most of the Gulfton residents are new to the U.S., about 90% speak little to no English, they lack transportation, and their unemployment rate is three times higher than the state level. This population includes approximately 10,200 children and youth, 3,000 of whom are less than five years old. Eighty percent of the elementary children have been categorized as at-risk of academic failure, 95% qualify for the free/reduced lunch program, and 75% have limited English skills. An additional 600 children are born each year, with about 25% of the women receiving late or no prenatal care (Gulfton Community Coalition, 1997). Gulfton is one of two Houston neighborhoods with the highest incidence of HIV/AIDS. While health care services are available, demand far exceeds supply. There are significant gaps in primary care, well-child, vision, and dental services. Special problems exist for those who are not certified for Medicaid and many minor health problems that go untreated in the early stages become costly and serious emergency room cases.

Apartment dwellers outnumber homeowners by far in this neighborhood. Apartment properties range in size from 20–1,800 units; Napoleon Square has, for example, 1,400 units. Caseworkers in the area report that it is not unusual to find as many as a dozen persons sharing the cost of a two-bedroom apartment. Many of these people are living day-to-day with much, if not all, of their time concentrated on providing the basic needs for their families. The entire Gulfton area suffers from a lack of clinics, parks, libraries, schools, and social services that these new residents so desperately need. Fortunately, work has begun to address some of the community needs. The Gulfton Community Coalition, for example, is working towards building a supportive environment for children. The Gulfton Youth Development Project offers a variety of programs for older youth by collaborating with organizations, businesses, churches, health clinics and hospitals, the police, and schools.

The BRIDGE/EL PUENTE was designed to address the needs of immigrants in the Gulfton Area who are constantly confronted by the many challenges of acculturation. The project developed partnerships in the Greater Gulfton Area among

Table 1. Agencies and Their Purposes.

Agency	Purpose
Neighborhood Centers Inc.	Serves as the fiscal agent for the BRIDGE/ EL PUENTE, as well as provides program coordination, child-development training for childcare volunteers, child-care services for the parents while attending program functions, and craft classes for the parents.
The University of Houston, Graduate School of Social Work, Office of Community Projects	Provides family support services.
The Texas Institute for the Arts in Education	Offers a seven-week session of parent training in learning to use performing arts activities to build basic skills for the preschool child.
The Houston Independent School District	(a) Provides home-based parent training for children ages 3 to 5 with weekly home visits year round, (b) facilitates school readiness as well as building and strengthening the parent-child relationship, and (c) links the parents to the school system.
The Houston Public Library	Offers four nine-week sessions to parents and children that develop the parent's role as first teacher, and promote the parent-child relationship through the use of books and homemade activities that emphasize concept building and language skills.
Memorial Hospital Southwest	Provides health education classes.
Napoleon Square Apartments	Provides the physical facility, easy access to residents, referrals and helps disseminate information to residents.
Christus Southwest Clinic	Provides preventive health services, prenatal care, well-baby checkups, immunizations and physicals for children age 0 to 6.
ESCAPE Family Resource Center	Conducts parenting workshops based on participants' needs and developmentally appropriate programs for children that focus on emotional competence and self-awareness.
GANO/CARECEN	Offers ESL and GED classes and promotes socialization for children age 0–6.
Healthy Families Initiatives	Conducts a comprehensive program for families with newborns, including: (a) a primary prevention program, (b) medical home link, (c) risk assessment of newborns and mothers, and (d) services in the home.

tenants, human service providers, academicians, and apartment managers, based upon the following goals: (a) the project's services will be integrated, and participants will be empowered through successful collaboration; (b) children completing the program will enter Kindergarten with skills and experiences that support lifelong learning and help them reach their fullest potential as learners; (c) parents of participants in the program will develop their knowledge of the normal developmental stages and the needs of children ages 0 to 6 years, and their mastery of parenting skills; (d) the program will demonstrate the business efficacy and profitability of organizing apartment complexes in the Greater Gulfton Area to serve as incubators of healthy children and economically viable families; and (e) preschool children and their mothers will find medical facilities that provide comprehensive care beginning, if possible, at preconception with special emphasis on comprehensive and preventive services for the 0 to 6 year-old-child and the pregnant mother.

There are several social service agencies committed to this project and they are briefly described in Table 1.

PARTICIPATORY EVALUATION PLAN FOR THE BRIDGE/EL PUENTE

An evaluation committee was formed by the GHCC several months before its projects were initially funded. The GHCC understood the importance of evaluation and the need to have external evaluators hired and working as soon as possible. They also understood the needs and concerns of evaluation as evidenced by their funding of the evaluation for six years, one entire year longer than the duration of the projects. The University of Houston Evaluation Team received the contract to conduct the external evaluation about two months after the project was initially funded.

The first six months of our external evaluation focused on developing a long-term evaluation plan. We initially met with several GHCC committees, including their evaluation committee. We also met with several committees from The BRIDGE/EL PUENTE, including their evaluation committee. It became clear from the outset that there were several different perspectives about the purpose of our external evaluation. Nearly everyone, however, was in favor of our proposed participatory evaluation approach because it was congruent with the "collaborative" focus of GHCC and BRIDGE/EL PUENTE. Because these projects focus on collaboration, it made sense that the evaluation should be collaborative or as it is often defined today "participatory" (Cousins & Whitmore, 1998). Participatory evaluation was thought to be an appropriate evaluation approach because we maintained that this would: (a) increase the utilization of results, (b) represent the multiple interests of stakeholder groups, and (c) empower the project staff.

Although there was a consensual agreement concerning the use of a participatory evaluation approach, there was much concern and debate about whether the evaluation should focus on the project's *processes* or *outcomes*. The chair of the GHCC evaluation committee was an experienced evaluator from a large foundation (that contributed money for the overall project and evaluation), who had been involved in several projects where he saw the futility of outcome-based evaluations in similar types of projects. Other GHCC committee members, who also represented large foundations, stressed the need to link project processes to specific, measurable outcomes. While this debate between process- and outcome-oriented evaluation still continues within the GHCC, at least up to now, the evaluation is still proceeding as if it is more process-based. It was determined, however, that the project would collect its own outcome data that would be part of its internal evaluation.

More specifically, our six-year external evaluation plan addresses the following six general areas: (a) evolution of consortia, (b) collaboration, (c) quality of services and practices, (d) program effectiveness, (e) efficiency and cost effectiveness, and (f) challenges and problems. Case study methods (Merriam, 1998) are used in the evaluation to primarily examine the project' processes. While most of the data collected in case study methods are qualitative (e.g. interviews and participant observations), we also supplement this "rich" data with quantitative data specifically collected by systematic observations of the parent training sessions and survey data from key individuals associated with the project. Our participatory approach to evaluation has allowed us to work collaboratively with the project leadership and staff in formulating and implementing the evaluation. We will describe this collaboration in more detail later in the chapter.

FUNDERS' PERSPECTIVES OF PROGRAM QUALITY

Initial Project Goals

To understand "quality" from the funders' vantage-point, one must first understand who they are and the reason they initiated the program in the first place. In 1994, representatives of 10 foundations initiated a series of meetings that ultimately led to the organization of the GHCC. First, they wanted to make a long-term difference in the lives of children and youth, and after considerable discussion, decided to focus on young children – birth through age six. Second, they had, and still have, an abiding faith that collaboration among agencies would strengthen services and improve the lives of young children and their parents. The processes of collaboration focuses on relationships among service providers, between service providers and the apartment complex residents, between funders and the project, and between the project and the community.

With the twin goals of focusing on young children and collaboration, GHCC issued a *Request for Proposal* in mid-1997, culled the pre-applications to seven, invited the seven to write full proposals, made site visits, and ultimately selected two proposals, one of which was The BRIDGE /EL PUENTE. Three outcomes dominated discussions of quality by GHCC leadership: (a) effectiveness of services, (b) efficiency of services, and (c) consequences of The BRIDGE/EL PUENTE. They are each described in the following sections.

Effectiveness of services. The relevant question about effectiveness was: *To what extent is collaboration among service providers strengthening the delivery of services?* The GHCC leadership wanted to determine if the funding it had put into the project was being effectively used: Were the agencies working together, supporting each other, recommending each other for specific activities, and building on residents' needs and aspirations? Effectiveness was judged in terms of the quality of collaboration among service providers. Their assumption was that more extensive collaboration would result in more effective service delivery. Indicators of effective collaboration included: (a) equality in the relationships among the participating groups, (b) a compelling plan for self-sufficiency, (c) integration of services, and (d) clearly identifiable outcomes that would be tracked over time.

Efficiency of services. The operable question here was: *To what extent is The BRIDGE/EL PUENTE a cost-effective enterprise?* GHCC foundation representatives did not operate in a vacuum, but were part of the larger organizations to which they were responsible. Their CEOs and governing boards expected them to monitor the quality of GHCC in terms of its efficiency of operation. Not everyone was convinced that collaboration would be cost-effective – or, in some cases, even effective.

Indicators of the project's efficiency, as conceived by the Executive Committee, included: (a) maximizes resources, and (b) streamlines services. This was stated elsewhere as a "seamless array of services rather than fragmented services now delivered for children".

Consequences of The BRIDGE/EL PUENTE. Consequence outcomes are defined and assessed not by the effectiveness nor the efficiency of services, not by what the various service providers do or the extensiveness of their services, but are based on the outcomes of these actions. Two consequence outcomes were inherent in GHCC leaders' conception of quality: (a) the impact of the project on the residents of Napoleon Square Apartments, and (b) the impact of the project on public policy and on the adoption by other agencies of The BRIDGE/EL PUENTE processes.

In assessing the first consequence outcome, the relevant question for GHCC was: *To what extent is The BRIDGE/EL PUENTE improving the lives of Hispanic, largely immigrant children, birth to age six, and their parents who live in an apartment complex in southwest Houston?* Indicators that this was occurring were specified in a description of The BRIDGE/EL PUENTE: "The project: (a) helps residents build community-oriented programs, (b) nurtures their children's physical development, (c) promotes social maturity, (d) enhances learning, and (e) provides preventive health care". These five specifications defined the consequences for which the project would be held responsible in terms of its mission to the residents of Napoleon Square Apartments.

The second consequence outcome involved the impact of The BRIDGE/EL PUENTE on other agencies and on the public. Two relevant questions were: (a) *To what extent do others in Houston and elsewhere identify The BRIDGE/EL PUENTE processes as harbingers of the future?* and (b) *To what extent has The BRIDGE/ EL PUENTE influenced public policy?*

To assess quality based on the first outcome, other agencies and local and state governments serving children would need to: "see [The BRIDGE/EL PUENTE processes as providing] a greater long-term return on their investments and, more importantly, that would truly improve the lives of children". The second outcome was assessed in terms of the extent to which the project's procedures and outcomes were known to policy makers and tended to influence their policies. They would be assessed in terms of the extent to which: (a) additional resources were provided for young children and their parents, (b) public policy was influenced, (c) laws or regulations were passed that favored collaboration and young children, and (d) BRIDGE/EL PUENTE-like programs were organized in other cities.

The GHCC leadership, composed primarily of funders, invested more than money in The BRIDGE/EL PUENTE. They invested their own perceptions of the most effective ways to improve society and ferment social reform. They had gone to their respective foundations and pressed for support of the new initiative. They became personally committed as well as professionally involved. Their perceptions of *quality* were shaped by their prior experience, the policies of their respective funding bodies, and by collaborative interactions among themselves and with other professionals in the community.

QUALITY PERSPECTIVES FROM THE SERVICE PROVIDERS AND FAMILIES

Initial Perspectives of the Project

From the outset, there was a small group of committed individuals from several different agencies who saw the GHCC grant as an opportunity to extend their work in the Gulfton area and create stronger bonds with the families. The individuals and the agencies they represented had prior experience working together which made it easier to join forces and write the grant. With just a few exceptions, the key collaborators have remained the same since the final proposal was funded.

A formal structure was also developed for the project that included operative guidelines regarding scope and authority, structure, officers, Executive Committee, and amendments. Although Neighborhood Centers Inc. was the designated fiscal agent, the Executive Committee consists of individuals representing all of the service providers. Project staff whom are housed at the Napoleon Square apartment complex also represent various service providers.

Perceptions of Quality From Service Providers and Families

In our initial evaluations of the project, we have found numerous instances where client perspectives of program quality changed as they have become more familiar with the programs. In the beginning, the families did not trust the service providers. They did not want to leave their children in the adjoining room while they attended a class because they did not know, and, therefore, could not trust the child-care workers. As their trust grew, the mothers began to ask for more child-care time. The workers were no longer considered outsiders; they were part of the group because several of the mothers volunteered in the child-care area and other mothers became part-time employees. Because the women knew one another from classes and social events, they no longer felt they were leaving their children with strangers.

Another example of program quality from the family or client perspective is the mothers' very high rating of the class offered by the Public Library, and their request that the program should have a second part. They did, however, point out that more control, more commitment, and better attendance should be a priority for the program. At first glance, this sounds a bit restrictive, but one must realize that space in the child-care area is limited. The number of persons enrolled is directly related to the number of children who require child care. If a mother of three enrolls and then does not faithfully attend, one or possibly two other mothers and three children could have attended. The parents implied that the service providers could institute a waiting list for those never enrolled, and another list for those previously enrolled but unable to attend and complete the course.

Scheduling is one area where service providers and client opinions often differ. Many mothers point out that those women with husbands and children cannot attend evening classes between three in the afternoon, or whenever school lets out, and seven in the evening when dinner is consumed, homework completed, and children bathed and put to bed. English classes are offered by one of the service providers twice weekly, but clients feel they would benefit from the reinforcement of daily classes. The same point was made regarding the GED Program. The BRIDGE/EL PUENTE does not offer any classes on the weekend, yet many clients would like to take advantage of those two days when child-care arrangements are easier, husbands have more time for themselves, and the home schedule is less regimented by school and work hours. On the other hand, service providers often find it difficult to hire teachers who are interested in teaching evenings and on weekends. Consequently, they are more likely to schedule classes when they have teaching staff available.

The majority of those clients who have taken the classes in the project would like to see a broader program array with new offerings, because they have already taken advantage of the limited opportunities provided to them. They have learned that one gains more than knowledge of English in an English class. For them, this is a place where they have started to meet their neighbors, share life experiences, and conceive of a community made up by them and their new friends. Therefore, they wish to continue to learn more in order to strengthen the community they have begun to build.

The Home Instruction Program for Preschool Youngsters (HIPPY) is another example of how a program can succeed beyond the original intentions of the service providers. HIPPY is a program provided by the local school district that teaches mothers how to prepare their preschool children for school by having a trained aide visit their home each week and show mothers how to use literacy materials with their children. If the mother is illiterate or semi-literate, the aide teaches the mother how to read and use the materials. The obvious benefits of this program are fairly explicit in that it is designed to improve the literacy skills of young children. Parents, however, have described to us other intangible benefits of the program. This program exemplifies to the families that public schooling in the United States is available to *everyone*, and that the public schools care enough to help prepare their children for school even though they do not speak English and may not be legal immigrants. Families have similar feelings about the Library program because many of them come from places in Latin America where libraries and books are only available for the wealthy and the educated. While the service providers generally see the library and HIPPY programs as merely literacy-enhancing activities, the families see them as empowering messages that give them hope for the future.

The above are instances of the perspectives held by service providers and clients. Others include the distance between what the service provider offers – a certain very limited program or course with specific goals and outcomes – and the client needs. If a client (a mother, in this case) gains enough confidence to go out by herself after building a trusting relationship with a child-care facility, it is only natural that she should desire an arrangement that would allow her to use the facility more often. She could, for example, leave her children at a child-care facility while tending to her needs (classes for self-improvement) or those of one of the children (health care). This, however, is not part of the child-care provider's commitment to the project.

Typically, service providers do not anticipate a change in the volume of their services as the project grows. They fashion programs as static entities, and funders favor these. While the newly trained family expects more, they get the same service they received at the beginning of the process. However, if a project is involved with a community in which they have to gain the trust of the clientele prior to their intervention, they must believe that these clients will develop, and be transformed, and as such must be prepared to meet their changing needs and desires.

SUMMARY AND DISCUSSION

This summary section is divided into three parts. The first part, Quality of Program Issues in Participatory Evaluation, focuses on issues evaluators need to address when conducting participatory evaluations. The second part, Buscando Su Voz en Dos Culturas (Finding Your Voice in Two Cultures), addresses the important evaluation issue of trying to understand the different worlds and cultures of the clients we are serving. The final part, Evaluation Challenges and Lessons Learned, focuses on the evaluation lessons we have learned from working on this project.

Quality of Program Issues in Participatory Evaluation

One of the real concerns involving the use of a participatory evaluation approach is that our close collaboration with the project puts us at risk of analyzing and interpreting data in terms of advocacy rather than in terms of accuracy. This advocacy stance, however, is not only common to participatory evaluation approaches. In practice, most evaluators attend to the interests of the party that pays for the evaluation rather than concerning themselves with the quality of the program they are evaluating (Stake, 1998). Nevertheless, many of the criticisms aimed at participatory evaluation focus on the potential for conflicts of interest and biased results because individuals who are often integrally involved in the program are also involved in the evaluation process of deciding issues such as what variables should be examined and how the data should be collected, analyzed, and reported (Brisolara, 1998; Smith, 1999).

There is not one correct method or way to conduct a participatory evaluation (Burke, 1998; Cousins & Whitmore, 1998; Whitmore, 1998). There are, however, some key elements of the process that are often included. The participatory evaluation process should:

(1) be participatory, with the key stakeholders actively involved in the decision making,
(2) acknowledge and address inequities of power and voice among participating stakeholders,
(3) be explicitly "political",
(4) use multiple and varied approaches to codify data,
(5) have an action component in order to be useful to the program's end users,
(6) explicitly aim to build capacity, especially evaluation capacity so that stakeholders can control future evaluation processes,
(7) be educational (Brisolara, 1998, pp. 45–46).

Most of these elements are often found in other evaluation approaches and none of them other than perhaps the first element of "participatory" seem inappropriate for evaluations of social service programs. Yet, it is the participatory aspect of the evaluation approach that makes the inquiry and findings relevant for the consumers and stakeholders. The participatory evaluation approach also aims at increasing the consumers and stakeholders' sense of empowerment. The participatory evaluation approach is practical in that it addresses the interests and concerns of consumers and it seeks to improve program outcomes rather than prove the value of the program. This is the key point to consider when addressing program quality issues.

Buscando Su Voz en Dos Culturas (Finding Your Voice in Two Cultures)

Finding one's voice in two cultures, two worlds, two societies, two ways of knowing, and two ways of being can be one of the most difficult challenges an individual faces

> today. For those who find themselves in a new land, the voice of life in another time, another generation, another country does not go away. However, those who relocate need to ask constantly: How much of ourselves do we have to give up in order to be accepted? How much of our way of life has to change in order for us to succeed, pursue happiness, or otherwise be comfortable in this "new and different" country (Cline, 1998, p. 701).

The preceding quote not only illustrates some of the problems facing the families of Napoleon Square, but it also exemplifies some of the perceived problems that could be related to program quality issues in the project. The BRIDGE/EL PUENTE is a complex project because it is trying to "bridge the gap" between the worlds and cultures of the people living in Napoleon Square versus the mainstream culture. The initial director of the project, for example, was often placed in awkward situations because she was the person primarily responsible for negotiating between the two worlds. During the first year, there were some concerns regarding internal communication in the project, but this issue was resolved through a few retreats that focused on involving everyone in the project (including evaluators) to address the dire needs and concerns of the families from Napoleon Square. Once everyone began to understand the serious problems and urgent needs of the families, there seemed to be a stronger commitment toward the success of the project from nearly everyone involved. This commitment towards the project's success helps continue to build bridges between the two cultures.

Stanfield (1999) warns evaluators that traditional social science frameworks hinder our ability to evaluate culturally and socially different worlds and realities. Participatory evaluation is especially suited for this project because it typically acknowledges and addresses inequities of power and voice among various stakeholders. Given that one of the goals of participatory evaluation is to empower those most affected by the program being reviewed (Brisolara, 1998), it is important that we try to obtain more input and involvement from the Napoleon Square community. Currently, very few apartment residents are involved in the leadership and organization of the project and none are involved in the evaluation. This is an area where improvements are necessary. A major part of this problem stems from the fact that all of the families involved in the project only speak Spanish, while the project meetings are generally conducted in English. Some meetings with service providers have recently been held in both Spanish and English, so a few project participants are finally starting to attend meetings. Language and culture are so intertwined that it is difficult to understand a culture without knowing the language. Both participants and project leaders need to expend greater effort to learn more about each other's language and culture.

Evaluation Challenges and Lessons Learned

Our evaluation of The BRIDGE/EL PUENTE has raised a number of challenges and provided us with some invaluable experiences during the first two years of the project. We plan to continue to work on several issues regarding understanding and practicing participatory evaluation. This includes more collaborative inquiry among the project service providers and the University of Houston Evaluation Team, as well as greater depth of participation. The monthly project evaluation meetings have provided a forum for such participation and pave the way for more extensive involvement in upcoming years. During the first year, this meeting consisted of the Project Director, Chair of The BRIDGE/EL PUENTE Evaluation Committee, one of the full-time service providers, and the University of Houston Evaluation Team. During the second year, this group expanded to include the Chair of The BRIDGE/EL PUENTE Governing Council and the Director of Early Childhood Education for Neighborhood Centers Inc. (i.e. the fiscal agent for the project). These two individuals are both key stakeholders and their genuine interest in the project and expertise in evaluation have made the meetings more substantive and policy oriented. They have illustrated how *we* can improve the evaluation process by changing the focus of the meetings and, thereby, improve the overall project.

All the first year evaluation meetings were held at the University of Houston, but in our efforts to become more collaborative, we decided to hold all Year 2 and Year 3 meetings at the Neighborhood Centers Inc. offices. This made the meetings more accessible for the project participants and sent an important message to the group that this was *their* evaluation, not just *ours*. The monthly meetings focused on both our external evaluation as well as their internal evaluation activities. We have examined a number of issues, such as selecting topics and questions to be addressed, instrument design and data collection, and some strategic planning. There appears to be "true" collaboration among all the participants at the meetings, but again there are clearly some areas where improvements need to be made. The University of Houston Evaluation Team, for example, wrote the entire evaluation report, while the rest of the group merely provided them with feedback. While it may not be feasible to expect the participants of the evaluation committee to contribute to the writing of the evaluation reports, they can perhaps be more involved in the collection of data, the interpretation of data, and the strategic planning for the evaluation. At one of our recent meetings, for example, a few of the project leaders suggested that some parents who work in the project's child-care center could help us collect evaluation data by interviewing other parents in the project. While this is only a modest involvement in the evaluation on the part of the families, it is nevertheless a start.

Another lesson we learned focuses on the types of data we have collected during the first two years of the evaluation. In our quest for collecting "objective" unbiased data, we may have focused too much on discrete aspects of the project we could measure. This is probably due to our social science positivist training as well as our underlying assumption that the funders of the project value such "objective" measures more than qualitative, rich data. In future years, we plan to focus more on experiential accounts, narratives, and stories to portray the project. Such data may do a better job of examining the quality of the project than the objective-type of data we have previously collected. The parent interviews we conducted during Year 2, for example, illustrate the richness and value of focusing on illuminative cases.

Other evaluation aspects we might examine in future years include exploring the impact of the participatory evaluation process. Following Papineau and Kiely (1996), we might interview participants involved in the evaluation process to respond to questions that address: (a) their actual involvement in the evaluation experience; (b) their attitudes toward the evaluation process, including perceived strengths, weaknesses, and things they would like to change; and (c) how the evaluation process has influenced them and the project. This type of study would help us assess the extent to which the evaluation process has been valuable to all the participants.

A final challenge that has recently arisen is that the evaluators and project leaders are facing some pressures from the funders who want the project to produce some "hard" outcomes soon, so they can continue to raise money to fund the project. Several of the key project leaders acknowledge the funders' position and accept that they will have to focus more on collecting outcome data, although they don't feel that it will necessarily improve the evaluation of the project. In our role as evaluators, we are also ambivalent about shifting our focus to a more outcomes-based evaluation. At this time, it is difficult to predict what will happen to the evaluation, but this conflict of perspectives has led to some positive, unanticipated outcomes. The funders, for example, have become much more involved in the evaluation process and the prospect of having regular evaluation meetings with project leaders, service providers, funders, and evaluators looks promising. Meanwhile, we will try to accelerate our efforts to include service providers and families in the evaluation process. While we clearly are not at the ideal collaborative stage of participatory evaluation, our efforts suggest that we are moving closer to that ideal model.

In conclusion, the value of participatory evaluation is that it allows stakeholders to be involved in the evaluation process. Clearly one of its major disadvantages is that it takes a lot of time. While we have had funders and project staff involved in the process, we still have not had the actual stakeholders of the project, the families, actively involved in the evaluation process. As we begin our third year of the evaluation, we are optimistic that our initial interviews with parents will help us sustain more meaningful relationships with them in following years and perhaps encourage them to become involved in the evaluation process. After all, if the project is to make a difference in the lives of the families and children living at Napoleon Square, we must thoroughly examine their perspectives so that we can determine a critical aspect of the "true" quality of the project. Our final challenge, however, may be to decide how much we need to rely on the "outside" voices of evaluators, project staff, service providers, and funders, versus these "inside" voices from the families of Napoleon Square. In other words, we need to be aware that program quality may be quite subjective, so we will need to continue to focus on quality issues from multiple perspectives.

The overall consequences of having different voices with divergent goals may be that the perceived overall quality of the program suffers because of pluralistic goals and purposes. On the other hand, the variety of perspectives may be quite valuable because it should reveal the flexibility as well as the complexity of the project. Rather than working on a consensus of program quality, we need to highlight and value the differences of perspectives inherent in this project. We will similarly need to effectively communicate and be respectful of the various perspectives of stakeholders in the program. In other words, the quality of The BRIDGE/EL PUENTE may only be determined once we understand the diversity of perceived quality among funders, service providers, and participants. Consequently, it is our role as external evaluators to portray the various perspectives of program quality.

REFERENCES

Brandon, P. R. (1998). Stakeholder participation for the purpose of helping to ensure evaluation validity: Bridging the gap between collaborative and non-collaborative evaluations. *American Journal of Evaluation*, *19*, 325–337.

Brisolara, S. (1998). The history of participatory evaluation and current debates in the field. In: E. Whitmore (Ed.), *Understanding and Practicing Participatory Evaluation* (pp. 25–41). San Francisco: Jossey-Bass.

Burke, B. (1998). Evaluating for a change: Reflections on participatory methodology. In: E. Whitmore (Ed.), *Understanding and Practicing Participatory Evaluation* (pp. 43–56). San Francisco: Jossey-Bass.

Cline, Z. (1998). Buscando su voz en dos culturas – Finding your voice in two cultures. *Phi Delta Kappan*, *79*, 699–702.

Cousins, J. B., & Earl, L. M. (1992). The case for participatory evaluation. *Educational Evaluation and Policy Analysis*, *14*, 397–418.

Cousins, J. B., & Earl, L. M. (1995a). Participatory evaluation enhancing evaluation use and organizational learning capacity. *The Evaluation Exchange: Emerging Strategies in Evaluating Child and Family Services*, *1*(3/4). Retrieved March 10, 1998 from the World Wide Web: http://gseweb.harvard.edu/~hfrp

Cousins, J. B., & Earl, L. M. (1995b). *Participatory evaluation in education: Studies in evaluation use and organizational learning*. London: Falmer.

Cousins, J. B., & Whitmore, E. (1998). Framing participatory evaluation. In: E. Whitmore, (Ed.), *Understanding and Practicing Participatory Evaluation* (pp. 5–23). San Francisco: Jossey-Bass.

Gulfton Community Coalition (1997). *The BRIDGE/EL PUENTE: Building resources internally for development and growth for everyone/Pueblos unidos en nuevas tareas educativas*. Houston, TX: Neighborhood Centers, Inc.

Merriam, S. B. (1998). *Qualitative research and case study applications in education*. San Francisco: Jossey-Bass.

Papineau, D., & Kiely, M. C. (1996). Participatory evaluation in a community organization: Fostering stakeholder empowerment and utilization. *Evaluation and Program Planning*, *19*, 79–93.

Smith, M. F. (1999). Participatory evaluation: Not working or not tested? *American Journal of Evaluation*, *20*, 295–308.

Stake, R. E. (1995). *The art of case study research*. Thousand Oaks, CA: Sage.

Stake, R. (1998). When policy is merely promotion: By what ethic lives an evaluator? *Studies in Educational Evaluation*, *24*, 203–212.

Stanfield, J. H. II. (1999). Slipping through the front door: Relevant social scientific evaluation in the people of color century. *American Journal of Evaluation*, *20*, 415–431.

Upshur, C. C., & Barreto-Cortez, E. (1995). What is participatory evaluation (PE) and what are its roots. *The Evaluation Exchange: Emerging Strategies in Evaluating Child and Family Services*, *1*(3/4). Retrieved March 10, 1998 from the World Wide Web: http://gseweb.harvard.edu/~hfrp

Whitmore, E. (Ed.) (1998). *Understanding and practicing participatory evaluation*. San Francisco: Jossey-Bass.

Whyte, W. F. (1989). Advancing scientific knowledge through participatory action research. *Sociological Forum*, *4*, 367–385.

Whyte, W. F. (1991). *Participatory action research*. Newbury Park, CA: Sage.

Worthen, B. R., Sanders, J. R., & Fitzpatrick, J. L. (1997). *Program evaluation: Alternative approaches and practical guidelines* (2nd ed.). New York: Longman.

SECTION III

11. USE AS A CRITERION OF QUALITY IN EVALUATION

Michael Quinn Patton

INTRODUCTION

Quality care. Quality education. Quality parenting. Quality time. Total quality management. Continuous quality improvement. Quality control. Quality assurance. Baldrige Quality Award. Quality is the watchword of our times. People in "knowledge-intensive societies . . . prefer 'better' to 'more' " (Cleveland, 1989, p. 157).

At the most fundamental level, the debates about abortion, on the one hand, and death with dignity and physician-assisted suicide, on the other, concern, in part, what is meant by "quality of life" (Humphrey, 1991). Kenneth E. Boulding (1985), one of the most prominent intellectuals of the modern era, devoted a book to the grand and grandiose topic of Human Betterment. In that book he defined development as "the learning of quality" (see Chap. 8). He struggled – ultimately in vain – to define "quality of life." He found the idea beyond determinative explication.

Then there is the day-to-day, more mundane side of quality of life. Quality has become the primary marketing theme of our age, e.g. "Quality is Job One" (Ford advertising slogan). Customers demand quality. This may stem, at least in part, from the fact that in the busy lives we now lead in Western society, we simply don't have time for things to break down. We don't have time to wait for repairs. We can't afford the lost productivity of not having things work (either products or programs). We have taken to heart the admonition that, in

Visions of Quality, Volume 7, pages 155–180.

ISBN: 0-7623-0771-4

the long run, it is cheaper do it right the first time – what ever "it" or "right" may refer to. "It" in this case refers to evaluation. Doing it "right" – now there's the rub. What does it mean – do an evaluation "right"? The answer takes us down the path to defining quality in relation to evaluation. Before taking on that specific issue, especially with regard to utilization-focused evaluation, it's worth spending just a bit more time reflecting on the meaning of quality in our times, for that is the context within which we must inevitably examine what quality means with reference to evaluation.

HISTORICAL AND CULTURAL CONTEXT

"Come, give us a taste of your quality" (Shakespeare, Hamlet, Act II, scene ii).

The current mania spot-lighting quality can give one the impression that this is a relatively recent concern, but the "fathers" of the quality movement – W. Edwards Deming and Joseph M. Juran – were preaching quality before World War II. In the 1930s, for example, Juran was applying concepts of empowered worker teams and continuous quality improvement to reduce defects at Western Electric's Chicago Hawthorne Works (Deutsch, 1998; Juran, 1951). Deming and his disciples have long viewed quality from the customer perspective, so quality is meeting or exceeding customer expectations. Twenty years ago Philip B. Crosby (1979) wrote a best-selling book on *The Art of Making Quality Certain*. Some of his assertions have become classic: "The first struggle, and it is never over, is to overcome the 'conventional wisdom regarding quality' " (p. 7); "The cost of quality is the expense of doing things wrong" (p. 11); "Quality is ballet, not hockey" (p. 13); "The problem of quality management is not what people don't know about. The problem is what they think they do know" (p. 13); "Quality management is a systematic way of guaranteeing that organized activities happen the way they are planned . . . Quality management is needed because nothing is simple anymore, if indeed it ever was" (p. 19).

Efforts to implement these and other principles swelled crescendo-like into a national mania as management consultants everywhere sang of total quality management and continuous quality improvement. The music began in the corporate sector, but by the early 1990s "the cult of total quality" had permeated deeply into the government and non-profit sectors (Walters, 1992). The Malcolm Baldrige National Quality Awards became, and remain, the pinnacle of recognition that the mountaintop of quality has been ascended.

Nor was concern about quality limited to management books. Robert Pirsig's (1974) classic *Zen and the Art of Motorcycle Maintenance* was an investigation into quality and excellence, themes he re-visited in *Lila: An Inquiry into Morals* (Pirsig, 1991) as he explored the "Metaphysics of Quality."

> What the Metaphysics of Quality adds to James' *pragmatism* and his *radical empiricism* is the idea that the primal reality from which subjects and objects spring is *value*. By doing so it seems to unite pragmatism and radical empiricism into a single fabric. Value, the pragmatic test of truth, is also the primary empirical experience. The metaphysics of quality says pure experience is value. Experience which is not valued is not experienced. The two are the same. This is where value fits. Value is not at the tail-end of a series of superficial scientific deductions that puts it somewhere in a mysterious undetermined location in the cortex of the brain. Value is at the very front of the empirical procession (Pirsig, 1991, p. 365; emphasis in the original).

Which would seem to bring us back to evaluation – making judgments of merit and worth, or *valuing*.

QUALITY ASSURANCE AND PROGRAM EVALUATION

Program evaluation and quality assurance have developed as separate functions with distinct purposes, largely separate literatures, different practitioners, and varying jargon. Each began quite narrowly but each has broadened its scope, purposes, methods, and applications to the point where there is a great deal of overlap and, most important, both functions can now be built on a single, comprehensive program information system.

Program evaluation traces its modern beginnings to the educational testing work of Thorndike and colleagues in the early 1900s. Program evaluation has traditionally been associated with measuring attainment of goals and objectives. The emphasis has been on finding out if a program "works," if it is effective. This has come to be called "summative" evaluation with a heavy reliance on experimental designs and quantitative measurement of outcomes.

Quality assurance (QA) in the United States had its official birth with the passage of the Community Mental Health Act Amendments of 1975.[1] This law required federally funded community mental health centers to have QA efforts with utilization and peer review systems. Its purpose was to assure funders, including insurers and consumers, that established standards of care were being provided.

Traditionally, given their different origins, program evaluation and quality assurance have had different emphases. These differences are summarized in Fig. 1.

---

1. Public Law 94-63.

Program Evaluation	Quality Assurance
1. Focus on program processes and outcomes	1. Focus on individual processes and outcomes
2. Aggregate data	2. Individual clinical cases
3. Goal-based judgment	3. Professional judgment-based
4. Intended for decision makers	4. Intended for clinical staff

Fig. 1. Comparing Program Evaluation and Quality Assurance.

The distinctions between QA and program evaluation have lost their importance as both functions have expanded. Program evaluation has come to pay much more attention to program processes, implementation issues, and qualitative data. Quality assurance has come to pay much more attention to outcomes, aggregate data, and cumulative information over time.

What has driven the developments in both program evaluation and quality assurance – and what now makes them more similar than different – are concerns about program improvement and gathering really useful information. Both functions had their origins in demands for accountability. Quality assurance began with a heavy emphasis on quality control but attention has shifted to concern for quality enhancement. Program evaluation began with an emphasis on summative judgments about whether a program was effective or not but has shifted to improving program effectiveness ("formative evaluation"). It is in their shared concern for gathering useful information to support program improvement that program evaluation and quality assurance now find common ground. Accountability demands can be well served, in part, by evidence that programs are improving.

Both accountability demands and program improvement concerns require comprehensive program information systems. We've learned that such systems should be designed with the direct involvement of intended users; that information systems should be focused on critical success factors (not data on every variable a software expert can dream up); that systems should be streamlined with utility in mind; and that program improvement systems benefit from both qualitative and quantitative information, both case and aggregate data. Indeed, harking back to the opening discussion about total quality management and continuous quality improvement, the systems that support such efforts began with a heavy emphasis on statistical process control and "objective" indicators of performance, but have come increasingly to value qualitative perspectives on quality. It turns out that one of the particular strengths of qualitative inquiry, strangely enough, is illuminating the nature and meaning of *quality* in particular contexts (Patton, 1990, pp.108–111).

The cutting edge for both program evaluation and quality assurance is designing genuinely useful and meaningful information systems for accountability *and* program improvement with a focus on excellence and ongoing learning rather than control and punishment. The best systems of program evaluation and quality assurance will use shared processes and data and will be designed using basic principles of organizational development, lifelong learning, and positive reinforcement.

THE STANDARDS: EVALUATION'S QUALITY CRITERIA

Evaluators evaluate programs. But, turnabout being fair play, it seems only fair for practitioners and decision makers to evaluate program evaluation and assess quality assurance. The standards for evaluation assert that they should do so (as should evaluators) using the criteria of utility, practicality, propriety, and accuracy (Joint Committee, 1994). These criteria include attention to an evaluation's timeliness, relevance, understandability, appropriate use, and appropriateness of methods. *In effect, the standards represent evaluation's quality criteria.*

Consider the utility standards that are "intended to ensure that an evaluation will serve the practical information needs of intended users" (Joint Committee, 1994). The implication is that, for example, a program evaluation or quality assurance initiative that is intended to improve a program but does not lead to documentable program improvements is of poor quality and, itself, in need of improvement. A summative evaluation aimed at supporting major decision-making, but not used in making decisions, is of poor quality – regardless of methodological rigor or academic acclaim.

Are evaluations really to be judged on their use? This is a matter of some controversy. Among those most opposed to use as a criterion of quality is Scriven (1991), who argues that the primary criterion for judging evaluations should be their truthfulness; he goes on to assert a conflict between serving truth and serving utility. There can be a tension between "truth tests" and "utility tests," to be sure, but the evidence is that those who commission evaluations want both (Weiss & Bucuvalas, 1980). In response, the evaluation profession has adopted both utility and accuracy standards as criteria of excellence, though that doesn't remove the tension that can emerge in trying to attain both utility and accuracy. Let me move, then, to what we've learned about how to handle such tensions in striving for overall evaluation quality in conducting utilization-focused evaluations. I'll then address some additional challenges to conducting high quality utilization-focused evaluations.

UTILIZATION-FOCUSED EVALUATION

Utilization-Focused Evaluation begins with the premise that evaluations should be judged by their utility and actual use; therefore, evaluators should facilitate the evaluation process and design any evaluation with careful consideration of how everything that is done, *from beginning to end*, will affect use. Nor is use an abstraction. Use concerns how real people in the real world apply evaluation findings and experience the evaluation process. Therefore, the *focus* in utilization-focused evaluation is on *intended use by intended users*.

In any evaluation there are many potential stakeholders and an array of possible uses. Utilization-focused evaluation requires moving from the general and abstract, i.e. possible audiences and potential uses, to the real and specific: actual primary intended users and their explicit commitments to concrete, specific uses. The evaluator facilitates judgment and decision-making by intended users rather than acting as a distant, independent judge. Because no evaluation can be value-free, utilization-focused evaluation answers the question of whose values will frame the evaluation by working with clearly identified, primary intended users who have responsibility to apply evaluation findings and implement recommendations. In essence, I argue, evaluation use is too important to be left to evaluators.

The utilization-focused approach is highly personal and situational. The evaluation facilitator develops a working relationship with intended users to help them determine what kind of evaluation they need. This requires negotiation in which the evaluator offers a menu of possibilities within the framework of established evaluation standards and principles. While concern about utility drives a utilization-focused evaluation, the evaluator must also attend to the evaluation's accuracy, feasibility and propriety (Joint Committee on Standards, 1994). Moreover, as a professional, the evaluator has a responsibility to act in accordance with the profession's adopted principles of conducting systematic, data-based inquiries; performing competently; ensuring the honesty and integrity of the entire evaluation process; respecting the people involved in, and affected by, the evaluation; and being sensitive to the diversity of interests and values that may be related to the general and public welfare (American Evaluation Association, 1995).

Utilization-focused evaluation does not advocate any particular evaluation content, model, method, theory or even use. Rather, it is a process for helping primary intended users select the most appropriate content, model, methods, theory and uses for their particular situation. Situational responsiveness guides the interactive process between evaluator and primary intended users. Many options are now available in the feast that has become the field of evaluation.

In considering the rich and varied menu of evaluation, utilization-focused evaluation can include any evaluative purpose (formative, summative, developmental), any kind of data (quantitative, qualitative, mixed), any kind of design (e.g. naturalistic, experimental) and any kind of focus (processes, outcomes, impacts, costs, and cost-benefit, among many possibilities).

> *Utilization-focused evaluation is a process for making decisions about these issues in collaboration with an identified group of primary users focusing on their intended uses of evaluation.*

A psychology of use undergirds and informs utilization-focused evaluation. In essence, research and my own experience indicate that intended users are more likely to use evaluations if they understand and feel ownership of the evaluation process and findings; they are more likely to understand and feel ownership if they've been actively involved; and by actively involving primary intended users, the evaluator is training users in use, preparing the groundwork for use, and reinforcing the intended utility of the evaluation every step along the way.

Quality Challenges in Conducting Utilization-Focused Evaluations

Given this brief overview of utilization-focused evaluation, what are the quality challenges involved in conducting such evaluations? They are these: (a) Selecting the "right" intended users; (b) working effectively with intended users, that is, attaining high quality interactions between intended users and the evaluator; (c) focusing the evaluation on intended uses; (d) maintaining technical quality to make sure the evaluation is worth using; (e) retaining independence of judgment and evaluation credibility as the evaluator builds close working relationships with intended users; (f) dealing with changes in intended users and uses; and (g) maintaining a strong sense of ethics and adhering to professional standards of conduct in the face of potential conflicts of interest.

Selecting the "Right" Intended Users: The Personal Factor

Many decisions must be made in any evaluation. The purpose of the evaluation must be determined. Concrete evaluative criteria for judging program success will usually have to be established. Methods will have to be selected and timelines agreed on. All of these are important issues in any evaluation. The question is: Who will decide these issues? The utilization-focused answer is: *primary intended users of the evaluation.*

Evaluation stakeholders are people who have a stake – a vested interest – in evaluation findings. For any evaluation there are multiple possible stakeholders: program funders, staff, administrators, and clients or program participants.

Others with a direct, or even indirect, interest in program effectiveness may be considered stakeholders, including journalists and members of the general public, or, more specifically, taxpayers, in the case of public programs. Stakeholders include anyone who makes decisions or desires information about a program. However, stakeholders typically have diverse and often competing interests. No evaluation can answer all potential questions equally well. This means that some process is necessary for narrowing the range of possible questions to focus the evaluation. In utilization-focused evaluation this process begins by narrowing the list of potential stakeholders to a much shorter, more specific group of primary *intended users*. Their information needs, i.e. their intended uses, focus the evaluation.

Different people see things differently and have varying interests and needs. I take that to be on the order of a truism. The point is that this truism is regularly and consistently ignored in the design of evaluation studies. To target an evaluation at the information needs of a specific person or a group of identifiable and interacting persons is quite different from what has been traditionally recommended as "identifying the audience" for an evaluation. Audiences are amorphous, anonymous entities. Nor is it sufficient to identify an agency or organization as a recipient of the evaluation report. Organizations are an impersonal collection of hierarchical positions. People, not organizations, use evaluation information – thus the importance of the personal factor:

> The personal factor is the presence of an identifiable individual or group of people who personally care about the evaluation and the findings it generates. Where such a person or group is present, evaluations are significantly more likely to be used; where the personal factor is absent, there is a correspondingly reduction in evaluation impact (Patton, 1997, p. 44).

The personal factor represents the leadership, interest, enthusiasm, determination, commitment, assertiveness, and caring of specific, individual people. These are people who actively seek information to make judgments and reduce decision uncertainties. They want to increase their ability to predict the outcomes of programmatic activity and thereby enhance their own discretion as decision makers, policy makers, consumers, program participants, funders, or whatever roles they play. These are the primary users of evaluation. Though the specifics vary from case to case, the pattern is markedly clear: Where the personal factor emerges, where some individuals take direct, personal responsibility for getting findings to the right people, evaluations have a greater chance of impact. Use is not simply determined by some configuration of abstract factors; it is determined in large part by real, live, caring human beings. "Nothing makes a larger difference in the use of evaluations than *the personal factor* [italics added] – the interest of officials in learning from the evaluation and the desire of the evaluator to get attention for what he knows" (Cronbach & associates, 1980,

p. 6). The importance of the personal factor in explaining and predicting evaluation use leads directly to the emphasis in utilization-focused evaluation on working with intended users to specify intended uses. The personal factor directs us to attend to specific people who understand, value and care about evaluation, and further directs us to attend to their interests. This is the primary lesson the profession has learned about enhancing use, and it is wisdom now widely acknowledged by practicing evaluators (Cousins & Earl, 1995).

Working Effectively With Intended Users: The Quality of Interactions

A common error I find among novice evaluators is thinking that the amount or quantity of stakeholder involvement is what constitutes quality. The quality of involvement, I've come to believe, is more important than quantity, though in some situations the two are clearly related. Facilitating high quality stakeholder involvement is the challenge of utilization-focused evaluation and implies considerable skill in working effectively with diverse intended users. It helps in this regard to be clear about the desired outcomes of stakeholder involvement and not let it become an end in itself.

Involving specific people who can and will use information enables them to establish direction for, commitment to, and ownership of the evaluation every step along the way from initiation of the study through the design and data collection stages right through to the final report and dissemination process. If intended users have shown little interest in the study in its earlier stages, my experience suggests that they are not likely to suddenly show an interest in using the findings at the end. They won't be sufficiently *prepared* for use. This preparation for use includes training intended users: teaching them to think empirically and evaluatively. Indeed, a major benefit of such training is the added value and impact of the evaluation beyond use of findings, what I've come to call "process use."

"Process use" refers to and is indicated by individual changes in thinking and behavior, and program or organizational changes in procedures and culture, that occur among those involved in evaluation as a result of the learning that occurs during the evaluation process. Evidence of process use is represented by the following kind of statement after an evaluation: "The impact on our program came not so much from the findings but from going through the thinking process that the evaluation required" (Patton, 1997, p. 90). (For an elaboration of process use, see Patton, 1997, Chap. 5.)

I use the phrase "active-reactive-adaptive" to suggest the nature of the consultative interactions that go on between evaluators and intended users. Utilization-focused evaluators are, first of all, active in deliberately and calculatedly identifying intended users and focusing useful questions. They are reactive in listening to intended users and responding to what they learn about the particular situation in which the evaluation unfolds. They are adaptive in altering evaluation questions and designs in light of their increased understanding of the situation and changing conditions. Active-reactive-adaptive evaluators don't impose cookbook designs. They don't do the same thing time after time. They are genuinely immersed in the challenges of each new setting and authentically responsive to the intended users of each new evaluation.

Working effectively with intended users to facilitate high quality interactions and involvement requires several evaluator skills that go well beyond the usual methods and technical training received in typical university training. These include skills in: (a) Communicating effectively (both orally and in writing) with lay people and non-academics; (b) facilitating group decision making and priority-setting processes; (c) clarifying values; (d) resolving and managing conflicts; (e) handling political conflicts; (f) being sensitive to divergent views, cultural differences, and individual values; (g) negotiating, including creating "win–win" outcomes in difficult negotiations; and (h) handling anger, aggression (both active and passive), and hidden agendas.

This list, which is far from exhaustive, makes it clear that high quality work with intended users requires a degree of skill and sophistication that goes well beyond methodological expertise. It also makes clear that not everyone is well-suited by temperament, training, or values to engage in utilization-focused evaluation, an understanding that, for me, has been deepened over the years through conversations with Robert Stake about the differences between responsive evaluation and utilization-focused evaluation.

RESPONSIVE EVALUATION AND UTILIZATION-FOCUSED EVALUATION

Responsive evaluation and utilization-focused evaluation are often lumped together as stakeholder-based approaches (Alkin, Hofstetter & Ai, 1998). Such a grouping together may disguise more than it clarifies. At any rate, a brief comparison will, I hope, illuminate the differences in quality criteria that often define different evaluation approaches or persuasions.

As noted earlier, a basic premise of utilization-focused evaluation is that no evaluation can serve all potential stakeholders' interests equally well. Utilization-focused evaluation makes explicit whose interests are served – those of specifically identified primary intended users. The evaluator then works with those primary intended users throughout the evaluation process. In contrast, responsive evaluation (Stake, 1975; Guba & Lincoln, 1981; Shadish, Cook & Leviton, 1995) "takes as its organizer the *concerns and issues of stakeholding audiences*" (Guba & Lincoln, 1981, p. 23). The evaluator interviews and observes stakeholders, then designs an evaluation that is responsive to stakeholders' issues. The evaluation report (or reports) would respond to the various stakeholder concerns identified by the evaluator initially. Stake's concerns about utilization-focused evaluation center on three primary issues: (a) threats to the evaluator's independence and, relatedly, insufficient notice that independence has been foregone; (b) overemphasis on the concerns of select clients thereby underserving larger audiences, and (c) elevation of utility as the primary criterion of merit over other dimensions of evaluation quality.

My critique of "responsive evaluation" is that evaluator independence is elevated above other criteria of quality, especially utility. Stakeholders are primarily sources of data and an audience for the evaluation, not real partners in the evaluation process. While responsive evaluators may get to know specific stakeholders fairly well during an evaluation, especially in collecting their experiences and perceptions, the evaluator maintains control over design and methods decisions. Responsive evaluation reports aim to provide the best possible description of a program and render an independent judgment of quality without catering to a particular set of identified users. My concern is that the audiences for such reports turn out to be relatively passive groups of largely anonymous faces: the "feds", state officials, the legislature, funders, clients, the program staff, the public, and so forth. If specific individuals are not identified from these audiences and organized in a manner that permits meaningful involvement in the evaluation process, then, by default, the evaluator becomes the real decision maker and stakeholder ownership suffers, with a corresponding threat to utility. In essence, responsive evaluation maintains the evaluator's control over the evaluation and avoids all the messiness, politics, and challenges of working directly with stakeholders. Such an approach presumes to maintain evaluator independence, not a small matter in my view, but at some risk to relevance and utility.

Stake worries that individuals chosen to represent stakeholder constituencies are rarely adequate representatives of those stakeholder groups. They may be imperfect representatives, to be sure, but the "responsive" alternative of having the evaluator represent stakeholders' views strikes me as even more problematic and, in the end, can dangerously feed researcher arrogance as the evaluator comes to believe that, because of superior technical expertise and the need for independence, he or she can determine what people ought to know better than they can decide themselves in direct negotiation and collaboration with the evaluator. I am unconvinced.

Let me add that, in my opinion, this difference in emphasis is not dependent on comparing external, summative evaluation (a common focus of Stake's work) with internal, developmental evaluation (a common focus of my work). While perceived evaluator independence is more critical to the credibility of external, summative evaluations, such evaluations, to be utilization-focused, still need to be directed to specific decision makers. To do so, the evaluator needs to work with those decision makers sufficiently to understand their information needs and evaluative criteria and design the evaluation accordingly. I have found that I can do this as an external, summative evaluator without compromising my independence, a point that critics dispute, but which is an empirical matter subject to independent assessment.

What should certainly be clear is that responsive evaluation and utilization-focused evaluation constitute significantly different approaches to stakeholders. Grouping them together as stakeholder-based approaches is problematic (Alkin, Hofsteter & Ai, 1998).

Focusing the Evaluation on Intended Uses

Utilization-focused evaluation aims to achieve intended use by intended users. Having examined some of the challenges to evaluation quality in identifying and working with intended users, it's time to examine the challenges to quality in focusing on intended uses. At the broadest level, evaluations serve one of three primary purposes: (1) rendering judgment about overall merit, worth or significance; (2) program improvement and quality enhancement; and/or (3) generating knowledge (Patton, 1997, Chap. 4). Chelimsky (1997) has called these the accountability, developmental, and knowledge perspectives, respectively. A common temptation on the part of both intended users and evaluators is to finesse the problem of focus by attempting to design an evaluation that serves all of these purposes. But, just as no single evaluation is likely to serve all intended users equally well, no single evaluation is likely to serve these competing purposes equally well. The utilization-focused evaluator bears the burden of facilitating choices from the rich and diverse menu of evaluation options. Different purposes will lead to different uses and will require evaluators to play different roles. Being clear about the tensions and trade-offs among purposes and uses helps focus an evaluation.

Gerald Barkdoll (1980), as associate commissioner for planning and evaluation of the U.S. Food and Drug Administration, identified three contrasting evaluator roles. Barkdoll found that the first role, "evaluator as scientist," was best fulfilled by aloof academics who focus on acquiring technically impeccable data while studiously staying above the fray of program politics and utilization relationships. The second role, which he called "consultative", was most suitable for evaluators who were comfortable operating in a collaborative style with policymakers and program analysts to develop consensus about their information needs and decide jointly the evaluation's design and uses. His third role, the "surveillance and compliance" evaluator, was characterized by aggressively independent and highly critical auditors committed to protecting the public interest and assuring accountability (e.g. Walters, 1992). These three roles reflect evaluation's historical development from three different traditions: (1) social science research, (2) pragmatic field practice, especially by internal evaluators and consultants, and (3) program and financial auditing.

When evaluation research aims to generate generalizable knowledge about causal linkages between a program intervention and outcomes, rigorous application of social science methods is called for and the evaluator's role as methodological expert will be primary. When the emphasis is on determining a program's overall merit or worth, the evaluator's role as judge takes center stage. If an evaluation has been commissioned because of, and is driven by, public accountability concerns, the evaluator's role as independent auditor, inspector, or investigator will be spotlighted for policymakers and the general public. When program improvement is the primary purpose, the evaluator plays an advisory and facilitative role with program staff. As a member of a design team, a developmental evaluator will play a consultative role. If an evaluation has a social justice agenda, the evaluator becomes a change agent.

In utilization-focused evaluation, the evaluator is always a negotiator – negotiating with primary intended users what other roles he or she will play. Beyond that, all roles are on the table, just as all methods are options. Role selection follows from and is dependent on sensitivity and attention to intended use by intended users.

Consider, for example, a national evaluation of Food Stamps to feed low-income families. For purposes of accountability and policy review, the primary intended users are members of the program's oversight committees in Congress (including staff to those committees). The program is highly visible, costly, and controversial, especially because special interest groups differ about its intended outcomes and who should be eligible. Under such conditions, the evaluation's credibility and utility will depend heavily on the perceived and actual independence, ideological neutrality, methodological expertise, and political savvy of the evaluators.

Contrast such a national accountability evaluation with an evaluator's role in helping a small, rural leadership program of the Cooperative Extension Service increase its impact. The program operates in a few local communities. The primary intended users are the county extension agents, elected county commissioners, and farmer representatives who have designed the program. Program improvement to increase participant satisfaction and behavior change is the intended purpose. Under these conditions, the evaluation's use will depend heavily on the evaluator's relationship with design team members. The evaluator will need to build a close, trusting, and mutually respectful relationship to effectively facilitate the team's decisions about evaluation priorities and methods of data collection, and then take them through a consensus-building process as results are interpreted and changes agreed on.

These contrasting case examples illustrate the range of contexts in which program evaluations occur. The precise role to be played in any particular study will depend on matching the evaluator's role with the context and purposes of the evaluation as negotiated with primary intended users. The quality of the evaluation depends on the goodness of fit between the evaluation's purpose, primary intended uses, and the evaluator's corresponding role in relation to purpose and intended uses. This means matching the evaluation to the situation.

SITUATIONAL EVALUATION

There is no one best way to conduct an evaluation. This insight is critical. The design of a particular evaluation depends on the people involved and their situation. A high quality evaluation, then, is one that is matched to the evaluation situation and context. The capacity to recognize situational variations and understand context can be added to our growing list of desirable evaluator characteristics for conducting high quality, utilization-focused evaluations.

Situational evaluation is like situation ethics (Fletcher, 1966), situational leadership (Blanchard, 1986; Hersey, 1985), or situated learning: "action is grounded in the concrete situation in which it occurs" (Anderson, Reder & Simon, 1996, p. 5). The standards and principles of evaluation provide overall direction, a foundation of ethical guidance, and a commitment to professional competence and integrity, but there are no absolute rules an evaluator can follow to know exactly what to do with specific users in a particular situation, thus the necessity of *negotiating* the evaluation's intended and expected uses.

Because every evaluation situation is unique, a successful evaluation (one that is useful, practical, ethical, and accurate) emerges from the special characteristics and conditions of the particular situation – a mixture of people, politics, history, context, resources, constraints, values, needs, interests, and chance. Despite the rather obvious, almost trite, and basically commonsense nature of this observation, it is not at all obvious to most stakeholders who worry a great deal about whether an evaluation is being done "right." Indeed, one common objection stakeholders make to getting actively involved in designing an evaluation is that they lack the knowledge to do it "right." The notion that there is one right way to do things dies hard. The right way, from a utilization-focused perspective, is the way that will be meaningful and useful to the specific evaluators and intended users involved, and finding that way requires interaction, negotiation, and situational analysis.

Utilization-focused evaluation is a problem-solving approach that calls for creative adaptation to changed and changing conditions, as opposed to a technical approach, which attempts to mold and define conditions to fit preconceived models of how things should be done. Narrow technocratic approaches emphasize following rules and standard operating procedures. Creative problem-solving approaches, in contrast, focus on what works and what makes sense in the situation. Standard methods recipe books aren't ignored. They just aren't taken as the final word. New ingredients are added to fit particular tastes. Homegrown or locally available ingredients replace the processed foods of the national supermarket chains, with the attendant risks of both greater failure and greater achievement.

USER RESPONSIVENESS AND TECHNICAL QUALITY

An evaluation's technical quality is related to appropriate use and is key to the question of whether an evaluation is worth using. One of the myths common in evaluation is that user responsiveness means some sacrifice of technical quality. Such an assertion implies that technical quality is judged by uniform and standardized criteria. But, as editors of scientific journals will tell you, peer reviewers rarely agree on their ratings of technical and methodological quality. In fact, standards of technical quality vary for different users and varying situations. It is not appropriate to apply the same technical standards in deciding to approve a prescription drug for widespread distribution and deciding whether to keep an after-school program open an hour later to be more responsive to parents. The issue, then, is not adherence to some absolute research standards of technical quality but, rather, making sure that methods and measures are *appropriate* to the validity and credibility needs of a particular evaluation purpose and specific intended users. This can involve trade-offs between traditional academic and scholarly standards of research quality (aimed at peer reviewed publication) and evaluation standards of methodological quality (aimed at credibility with primary stakeholders and intended users).

Stake recognized this trade-off in conceptualizing responsive evaluation and making qualitative inquiry and case studies a foundation of responsiveness, emphasizing the strength of such methods to capture, portray, and respond to different stakeholder perspectives as well as to penetrate the inner workings of a program.

> Responsive evaluation is an alternative, an old alternative, based on what people do naturally to evaluate things; they observe and react. The approach is not new. But this alternative

> has been avoided in district, state, and federal planning documents and regulations because it is subjective and poorly suited to formal contracts. It is also capable of raising embarrassing questions (Stake, 1975, p. 14).

Stake (1975) also connected responsive evaluation to use in commenting that "it is an approach that trades off some measurement precision in order to increase the usefulness of the findings to persons in and around the program" (p. 14). Stake influenced a new generation of evaluators to think about the connection between methods and use, and his book on *The Art of Case Study Research* (1995), published two decades later, has already extended that influence. Case research has the virtue of being especially user friendly to non-academic and community-based users.

Jennifer Greene (1990) examined in depth the debate about "technical quality versus user responsiveness." She found general agreement that both are important, but disagreements about the relative priority of each. She concluded that the debate is really about how much to recognize and deal with evaluation's political inherency:

> Evaluators should recognize that tension and conflict in evaluation practice are virtually inevitable, that the demands imposed by most if not all definitions of responsiveness and technical quality (not to mention feasibility and propriety) will characteristically reflect the competing politics and values of the setting (p. 273).

She then recommended that evaluators "explicate the politics and values" that undergird decisions about purpose, audience, design, and methods. Her recommendation is consistent with utilization-focused evaluation.

Evaluation Credibility And Evaluator Independence

A common concern about utilization-focused evaluation is loss of evaluator independence and a corresponding loss of evaluation credibility as the evaluator builds close working relationships with intended users. For example, I encounter a lot of concern that in facilitating utilization-focused evaluation, the evaluator may become co-opted by stakeholders. How can evaluators maintain their integrity if they become involved in close, collaborative relationships with stakeholders? How does the evaluator take politics into account without becoming a political tool of only one partisan interest?

The nature of the relationship between evaluators and the people with whom they work is a complex and controversial one. On the one hand, evaluators are urged to maintain a respectful distance from the people they study to safeguard objectivity and minimize personal and political bias. On the other hand, the human relations perspective emphasizes that close, interpersonal contact is a necessary condition for building mutual understanding. Evaluators thus find themselves on the proverbial horns of a dilemma: getting too close to decision makers may jeopardize scientific credibility; remaining distant may undermine use.

A program auditor at a workshop put the issue less delicately when he asked, "How can we get in bed with decision makers without losing our virginity?" This is a fascinating and revealing metaphor, showing just how high the stakes can seem. The evaluator is portrayed as the innocent, the policy maker as the co-opting tempter planning a seduction. I once reported this metaphor to a group of policy-makers who immediately reframed the question: How do we get in bed with evaluators without getting sadistically abused?" Different stakes, different fears.

One way to handle concerns about the evaluator being co-opted is to stay focused on evaluation's empirical foundation, which includes making assumptions and values explicit, testing the validity of assumptions, and carefully examining a program to find out what is actually occurring. The integrity of an evaluation depends on its empirical orientation – that is, its commitment to systematic and credible data collection and reporting. Likewise, the integrity of an evaluation group process depends on helping participants adopt an empirical perspective. A commitment must be engendered to really find out what is happening, at least as nearly as one can given the limitations of research methods and scarce resources. Engendering such commitment involves teaching and facilitating.

When stakeholders first begin discussing the purpose of an evaluation they will often do so in non-empirical terms: "We want to prove the program's effectiveness". Proving effectiveness is a public relations job, not an evaluation task. This statement tells the evaluator about that person's attitude toward the program and it indicates a need for diplomatically, sensitively, but determinedly, reorienting that stakeholder from a concern with public relations to a concern with learning about and documenting actual program activities and effects.

Sometimes the opposite bias is the problem. Someone is determined to kill a program, to present only negative findings and to "prove" that the program is ineffective. In such cases the evaluator can emphasize what can be learned by finding out about the program's strengths. Few programs are complete disasters. An empirical approach means gathering data on *actual* program activities and effects, and then presenting those data in a fair and balanced way so that information-users and decision makers can make their own judgments about goodness or badness, effectiveness or ineffectiveness. Such judgments, however, are separate from the data. In my experience, most evaluation task force members will readily move into this kind of empirical orientation as they come to understand its utility and fairness. It's the evaluator's job to help them adopt and maintain, even deepen, that perspective through ongoing reinforcement.

Moreover, some stakeholders never make the shift. The savvy evaluator will monitor the empirical orientation of intended users and, in an active-reactive-adaptive mode of situational responsiveness, take appropriate steps to keep the evaluation on an empirical and useful path.

THE ETHICS OF BEING USER-FOCUSED

Concern that utilization-focused evaluators may be co-opted by stakeholders, or become pawns in service of their political agendas, raises questions beyond how to be politically astute, strategic and savvy, or how to prevent misuse. Underneath, decisions about one's relationships with intended users involve ethics. Speaking truth to power is risky – risky *business*. Not only is power involved, but money is involved. The Golden Rule of consulting is, "Know who has the gold". Evaluators work for paying clients. The jobs of internal evaluators depend on the pleasure of superiors and future work for independent evaluators depends on client satisfaction. Thus, there's always the fear that "they who pay the piper call the tune", meaning not just determining the focus of the evaluation, but also prescribing the results. Evaluators can find themselves in conflict between their professional commitment to honest reporting and their personal interest in monetary gain or having future work.

The Program Evaluation Standards (Joint Committee, 1994) provide general ethical guidance and the "Guiding Principles" of the American Evaluation Association insist on integrity and honesty throughout the entire evaluation process. Beyond general ethical sensitivity, however, the ethics of utilization-focused evaluators are most likely to be called into question around two essential aspects of utilization-focused evaluation: (a) limiting stakeholder involvement to primary intended users, and (b) working closely with those users. The ethics of limiting and focusing stakeholder involvement concerns who has access to the power of evaluation knowledge. The ethics of building close relationships concerns the integrity, neutrality and corruptibility of the evaluator. Both of these concerns center on the fundamental ethical question: Who does the evaluation – and an evaluator – serve?

I would offer three admonitions with regard to utilization-focused evaluation: (a) evaluators need to be deliberative and intentional about their own moral groundings; (b) evaluators must exercise care, including ethical care, in selecting projects to work on and stakeholders to work with; and (c) evaluators must be clear about whose interests are more and less represented in an evaluation. Let me elaborate these points.

First, evaluators need to be deliberative and intentional about their own moral groundings. The evaluator is also a stakeholder – not the primary stakeholder

– but, *in every evaluation, an evaluator's reputation, credibility and beliefs are on the line*. A utilization-focused evaluator is not passive in simply accepting and buying into whatever an intended user initially desires. The active-reactive-adaptive process of negotiating connotes an obligation on the part of the evaluator to represent the standards and principles of the profession as well as his or her own sense of morality and integrity, while also attending to and respecting the beliefs and concerns of other primary users.

The second point concerns project and client selection. One way in which I take into account the importance of the personal factor is by careful attention to who I work with. Whether one has much choice in that, or not, will affect the way in which ethical issues are addressed, especially what kinds of ethical issues are likely to be of concern. In challenging what he has called "clientism" – "the claim that whatever the client wants . . . is ethically correct", House (1995) asked: "What if the client is Adolph Eichmann, and he wants the evaluator to increase the efficiency of his concentration camps?" (p. 29). Clearly, evaluators must have a strong sense of what and whom they will and will not support.

I used the phrase "active-reactive-adaptive" to suggest the nature of the consultative interactions that go on between evaluators and intended users. Being active-reactive-adaptive explicitly recognizes the importance of the individual evaluator's experience, orientation, values, and contribution by placing the mandate to be "active" first in this consulting triangle. Situational responsiveness does not mean rolling over and playing dead (or passive) in the face of stakeholder interests or perceived needs. Just as the evaluator in utilization-focused evaluation does not unilaterally impose a focus and set of methods on a program, so, too, the stakeholders are not set up to impose their initial predilections unilaterally or dogmatically. Arriving at the final evaluation design is a negotiated process that allows the values and capabilities of the evaluator to intermingle with those of intended users.

A third issue concerns how the interests of various stakeholder groups are represented in a utilization-focused process. I'm reluctant, as a white, middle class male, to pretend to represent the interests of people of color or society's disadvantaged. My preferred solution is to work to get participants in affected groups representing themselves as part of the evaluation negotiating process. As discussed in distinguishing utilization-focused evaluation from responsive evaluation, being user-focused involves real people in an active process, not just attention to vague, abstract audiences. Thus, where the interests of disadvantaged people are at stake, ways of hearing from or involving them directly should be explored, not just have them represented in a potentially patronizing manner by the advantaged. Whether and how to do this may be part of what the evaluator attends to during active-reactive-adaptive interactions and negotiations.

Guarding Against Corruption of an Evaluation

Another concern about utilization-focused evaluation is that it fails to deliver on evaluation's central purpose: rendering independent judgments about merit or worth. If evaluators take on roles beyond judging merit or worth, like creating learning organizations or empowering participants, or, alternatively, eschew rendering judgment in order to facilitate judgments by intended users, has the evaluator abandoned his or her primary responsibility of rendering independent judgment?

For Scriven (1991), evaluators don't serve specific people. They serve truth. Truth may be a victim when evaluators form close working relationships with program staff. Scriven admonishes evaluators to guard their independence scrupulously. Involving intended users would only risk weakening the hard-hitting judgments the evaluator must render. Evaluators, he has observed, must be able to deal with the loneliness that may accompany independence and guard against "going native", the tendency to be co-opted by and become an advocate for the program being evaluated (p. 182). Going native leads to "incestuous relations" in which the "evaluator is 'in bed' with the program being evaluated" (p. 192). He has condemned any failure to render independent judgment as "the abrogation of the professional responsibility of the evaluator" (p. 32). He has derided what he mockingly called "a kinder, gentler approach" to evaluation (p. 39). His concerns stem from what he has experienced as the resistance of evaluation clients to negative findings and the difficulty evaluators have – psychologically – providing negative feedback. Thus, he has admonished evaluators to be uncompromising in reporting negative results. "The main reason that evaluators avoid negative conclusions is that they haven't the courage for it" (p. 42).

My experience has been different from Scriven's, so I reach different conclusions. Operating selectively, as I acknowledged earlier, I choose to work with clients who are hungry for quality information to improve programs. They are people of great competence and integrity who are able to use and balance both positive and negative information to make informed decisions. I take it as part of my responsibility to work with them in ways that they can hear the results, both positive and negative, and use them for intended purposes. I don't find them resistant. I find them quite eager to get quality information that they can use to develop the programs to which they have dedicated their energies. I try to render judgments, when we have negotiated my taking that role, in ways that can be heard, and I work with intended users to facilitate their arriving at their own conclusions. They are often harsher on themselves than I would be.

In my experience, it doesn't so much require courage to provide negative feedback as it requires skill. Nor do evaluation clients have to be unusually enlightened for negative feedback to be heard and used if, through skilled facilitation, the evaluator has built a foundation for such feedback so that it is welcomed for long-term effectiveness. Dedicated program staff don't want to waste their time doing things that don't work.

I have followed in the tracks of, and cleaned up the messes left by, evaluators who took pride in their courageous, hard-hitting negative feedback. They patted themselves on the back for their virtues and went away complaining about program resistance and hostility. I watched them in action. They were arrogant, insensitive and utterly unskilled in facilitating feedback as a learning experience. They congratulated themselves on their independence of judgment and commitment to "telling it like it is", and ignored their largely alienating and useless practices. They were closed to feedback about the ineffectiveness of their feedback.

It is from these kinds of experiences that I have developed a preference for constructive and utilization-focused feedback. In any form of feedback, it's hard to hear the substance when the tone is highly judgmental and demeaning. This applies to interactions between parents and children (in either direction), between lovers and spouses, among colleagues, and, most decidedly, between evaluators and intended users. Being kinder and gentler in an effort to be heard need not indicate a loss of virtue or cowardice. In this world of ever-greater diversity, sensitivity and respect are not only virtues, they're more effective and, in evaluation, more likely to lead to findings being heard, understood and actually used.

The fundamental value-premise of utilization-focused evaluation is that intended users are in the best position to decide for themselves what questions and inquiries are most important. From this perspective, moral inquiries and social justice concerns ought to be on the menu, not as greater goods, but as viable choices. Of course, what intended users choose to investigate will be determined by how they are chosen, who they are, what they represent and how the evaluator chooses to work with them – all decisions that involve both politics and ethics.

Standards and principles provide guidance for dealing with politics and ethics, but there are no absolute rules. These arenas of existential action are replete with dilemmas, conundrums, paradoxes, perplexities, quandaries, temptations and competing goods. Utilization-focused evaluation may well exacerbate such challenges, so warnings about potential political corruption, ethical entrapments and moral turpitude direct us to keep asking fundamental questions: What does it mean to be useful? Useful to whom? Who benefits? Who is hurt? Who decides? What values inform the selection of intended users and intended uses? Why? Or, why not? Quality depends on authentic engagement with, and answers to, these questions.

DEALING WITH CHANGES IN INTENDED USERS AND USES

The Achilles heel of Utilization-Focused Evaluation, its point of greatest vulnerability, is turnover of primary intended users. The quality of the process so depends on the active engagement of intended users that to lose users along the way to job transitions, reorganizations, reassignments and elections can undermine eventual use. Replacement users who join the evaluation late in the process seldom come with the same agenda as those who were present at the beginning. The best antidote involves working with a task force of multiple intended users so that the departure of one or two is less critical. Still, when substantial turnover of primary intended users occurs, it may be necessary to reignite the process by renegotiating the design and use commitments with the new arrivals on the scene.

Many challenges exist in selecting the right stakeholders, getting them to commit time and attention to the evaluation, dealing with political dynamics, building credibility and conducting the evaluation in an ethical manner. All of these challenges revolve around the relationship between the evaluator and intended users. When new intended users replace those who depart, new relationships must be built. That may mean delays in original timelines, but such delays pay off in eventual use by attending to the foundation of understandings and relationships upon which high quality utilization-focused evaluation is built.

The utilization-focused evaluation is seldom simple or linear. More commonly, the process of negotiating is circular and iterative as intended users are identified and re-identified, evaluation questions are focused and refocused, and methods are considered and reconsidered. The active-reactive-adaptive evaluator who is situationally responsive and politically sensitive may find that new stakeholders become important or new questions emerge in the midst of methods decisions. Nor is there a clear and clean distinction between the processes of focusing evaluation questions and making methods decisions.

Of course, the point of all this attention to the quality of the evaluation process is the hope that quality evaluations will improve the quality of programs, or at least improve the quality of decision making about programs such that better decisions lead to higher quality programs. It would seem appropriate, then, to comment briefly on what constitutes quality programming from an evaluation perspective.

PROGRAM QUALITY

Just as the quality movement in the business sector has focused on meeting or exceeding customer expectations, so program quality in the education and non-profit sectors begins with understanding the perspectives of program participants. Whether participants are called students, clients, customers, recipients, beneficiaries, or by some other nomenclature, their experiences and perspectives are the centerpiece of both high quality programming and high quality evaluation. By making the perspectives and experiences of participants the centerpiece of evaluative feedback and judgment, evaluators help represent and reinforce the message that program quality ultimately depends on the quality of participants' experiences. This is consistent with Schorr's (1988) "lessons" from "successful programs." She found that successful programs "all take their shape from the needs of those they serve rather than from the precepts, demands, and boundaries set by professionalism and bureaucracies" (p. 259). In other words, they are participant-centered and driven by participants' needs and reactions. In so doing, they are also results-focused.

This illuminates the connection between evaluation quality and program quality. I have emphasized the importance for evaluation of being user-focused. Intended users are the clients and participants in evaluation. Utilization-focused evaluation is user-centered. It is also results-driven in attending to intended uses by intended users, just as programs are outcomes-focused by attending to the real needs of those they aim to serve.

Evaluation quality and program quality are even more directly linked in that, from an evaluation perspective (at least my own), a high quality program will be one that is built on and developed by using the logic and principles of evaluative thinking. Evaluators contribute to program quality, then, not just by providing findings, but by supporting the use of evaluative logic in making program design and implementation decisions, what I referred to earlier as "process use" (Patton, 1997, Chap. 5). In that regard, then, a high quality program will manifest the following (as examples): (a) Clarity about goals, purposes and program priorities; (b) specificity about target populations and outcomes; (c) work planned systematically; (d) assumptions made explicit; (e) program concepts, ideas and goals operationalized; (f) inputs and processes distinguished from outcomes; (g) decisions and directions supported with empirical evidence and logic; (h) data-based statements of fact separated from interpretations and judgments; (i) criteria of quality and standards of excellence made explicit; (j) claims of success, or generalizations and causal explanations about program effectiveness limited to what data support; and (k) ongoing improvement based on evaluation.

I've become convinced that a common source of misunderstanding, even conflict, between evaluators and program staff is the failure for either to appreciate the extent to which evaluative thinking, logic, and values, as manifest in the preceding list, constitute an implicit set of criteria for judging the quality of program design and decision making, and ultimately, the quality of the program itself. For evaluators, these criteria spill over into other areas of life, as I discovered in reflecting on my experiences as a parent (Patton, 1999). These criteria and their potential operationalization in program processes are a large part of what evaluators offer programs in their efforts to enhance quality. This way of thinking about programs can be as eye-opening and change-provoking as actual evaluation findings.

QUALITY IN THE EYE OF THE INTENDED USER

I opened by describing the pervasiveness of the concern with quality in modern society. That provided the context for looking at quality in evaluation generally and quality in utilization-focused evaluation specifically. In essence, I have argued for utility as a primary criterion of evaluation quality. Utility is defined situationally as intended use by intended users, meaning that *evaluation quality ultimately resides in the perceptions, information needs, and actual uses of evaluation made by intended users*. Quality in this sense is not a property of the evaluation itself, but rather derives from the experiences of people with the evaluation – in the case of utilization-focused evaluation, the experiences of primary intended users. No matter how rigorous an evaluation's methods and no matter how independent the evaluator, if the resulting process and/or findings aren't perceived as useful and actually used by intended users in intended ways, I judge the overall evaluation to have been of poor quality.

Of course, other evaluators will emphasize different criteria of quality. For example, I find a great many evaluators clinging to their independence and arguing for the primacy of their own notions of what constitutes rigor despite evidence that their work is of little use to anyone other them themselves. They seem to believe that the best one can do is publish one's findings in the vague hope that someone (as yet unknown and unspecified) may happen upon them (at some unknown and unspecified future time) and become enlightened.

As this highly biased, even jaundiced, analysis shows, when engaging in dialogues about quality, we must deal with both values (our criteria of quality) and evidence (what actions meet or do not meet those criteria). Through such professional dialogues we can hope to contribute to improvements in both program and evaluation quality. That assumes, of course, that one is interested in such improvements. Some, perhaps many, may not be. So, after so many pages devoted to enhancing quality, and in the spirit of balance and even-handedness for which professional evaluators are rightly renowned, I close with an alternative perspective. Those sick and tired of all this concern about quality can avoid quality improvements by adhering religiously to the sage advice of comedian George Carlin:

> Never give up an idea simply because it is bad and doesn't work. Cling to it even when it is hopeless. Anyone can cut and run, but it takes a special person to stay with something that is stupid and harmful. . . . If by chance you make a fresh mistake, especially a costly one, try to repeat it a few times so you become familiar with it and can do it easily in the future. Write it down. Save it (Carlin, 1997, p. 218).

REFERENCES

Alkin, M., Hofstetter, C., & Ai, X. (1998). Stakeholder concepts in program evaluation. In: A. Reynolds & H. Walberg (Eds), *Advances in Educational Productivity* (pp. 87–113). Stamford, CT: JAI.

American Evaluation Association Task Force on Guiding Principles for Evaluators (1995, Summer). Guiding principles for evaluators. *New Directions for Program Evaluation, 66*, 19–34.

Anderson, J. R., Reder, L. M., & Simon, H. (1996). Situated learning and education. *Educational Researcher, 25*(4), 5–21.

Barkdoll, G. L. (1980). Type III evaluations: Consultation and consensus. *Public Administration Review, 2*, 174–179.

Blanchard, K. (Speaker) (1986). *Situational leadership* [Cassette Recording]. Escondido, CA: Blanchard Training and Development, Inc.

Boulding, K. E. (1985). *Human betterment.* Beverly Hills, CA: Sage.

Carlin, G. (1997). *Braindroppings.* New York: Hyperion.

Chelimsky, E. (1997). The coming transformations in evaluation. In: E. Chelimsky & W. R. Shadish (Eds), *Evaluation for the 21st Century* (pp. 1–26). Newbury Park, CA: Sage.

Cleveland, H. (1989). *The knowledge executive: Leadership in an information society.* New York: E. P. Dutton.

Cousins, J. B., & Earl, L. M. (Eds) (1995). *Participatory evaluation in education: Studies in evaluation use and organizational learning.* London: Falmer.

Cronbach, L. J. & associates. (1980). *Toward reform of program evaluation.* San Francisco: Jossey-Bass.

Crosby, P. B. (1979). *Quality is free: The art of making quality certain.* New York: McGraw-Hill.

Deutsch, C. H. (1998, November 15). The guru of doing it right still sees much work to do. *The New York Times*, p. BU-5.

Fletcher, J. (1966). *Situation ethics: The new morality.* London: Westminster John Knox.

Greene, J. C. (1990). Technical quality versus user responsiveness in evaluation practice. *Evaluation and Program Planning, 13*(3), 267–274.

Guba, E. G., & Lincoln, Y. S. (1981). *Effective evaluation: Improving the usefulness of evaluation results through responsive and naturalistic approaches.* San Francisco: Jossey-Bass.

Hersey, P. (1985). *Situational leader.* Charlotte, NC: Center for Leadership.

House, E. R. (1995). Principled evaluation: A critique of the AEA guiding principles. In: W. R. Shadish, D. L. Newman, M. A. Scheirer & C. Wye (Eds), *New Directions for Program*

Evaluation: Number 66. Guiding Principles for Evaluators (pp. 27–34). San Francisco: Jossey-Bass.

Humphrey, D. (1991). *Final exit*. Eugene, Oregon: The Hemlock Society.

Joint Committee on Standards for Educational Evaluation (1994). *The program evaluation standards*. Thousand Oaks, CA: Sage.

Juran, J. M. (1951). *Quality control handbook*. New York: McGraw-Hill.

Patton, M. Q. (1990). *Qualitative evaluation and research methods*. Newbury Park, CA: Sage.

Patton, M. Q. (1997). *Utilization-focused evaluation: The new century text*. Newbury Park, CA: Sage.

Patton, M. Q. (1999). *Grand Canyon celebration: A father-son journey of discovery*. Amherst, NY: Prometheus Books.

Pirsig, R. M. (1974). *Zen and the art of motorcycle maintenance*. New York: William Morrow and Company.

Pirsig, R. M. (1991). *Lila: An inquiry into morals*. New York: Bantam Books.

Schorr, L. B. (1988). *Within our reach: Breaking the cycle of disadvantage*. New York: Doubleday.

Scriven, M. (1991). Beyond formative and summative evaluation. In: M. W. McLaughlin & D. C. Phillips (Eds), *Evaluation and Education: At Quarter Century* (pp. 18–64). Chicago: University of Chicago Press.

Shadish, W. R., Cook, T. D., & Leviton, L. C. (1995). *Foundations of program evaluation: Theories in practice*. Newbury Park, CA: Sage.

Stake, R. E. (1975). *Evaluating the arts in education: A responsive approach*. Columbus, OH: Charles E. Merrill.

Stake, R. E. (1995). *The art of case study research*. Newbury Park, CA: Sage.

Walters, J. (1992, May). The cult of total quality. *Governing: The Magazine of States and Localities*, 38–42.

Weiss, C. H., & Bucuvalas, M. (1980). Truth tests and utility tests: Decision makers' frame of reference for social science research. *American Sociological Review*, *45*, 302–313.

12. DESCRIPTIVE VALUES AND SOCIAL JUSTICE

William R. Shadish and Laura C. Leviton

INTRODUCTION

> The human spirit glows from that small inner light of doubt whether we are right, while those who believe with complete certainty that they possess the right are dark inside . . . Those who enshrine the poor . . . are as guilty as other dogmatists and just as dangerous. To diminish the danger that ideology will deteriorate into dogma . . . no ideology should be more specific than that of America's founding fathers: "For the general welfare" (Alinsky, 1989, p. 4).

Many dictionaries suggest that quality can be defined as "degree of excellence". If so, then quality is closely tied to evaluation. After all, evaluation is a process for forming judgments about quality, and so that process embodies a theory of what quality is, and how it is to be judged. In this chapter, we address one part of that process, the role that the values of stakeholders should play in reaching quality judgments in evaluation. We especially argue that a socially just evaluation in a democracy is one that gives voice to all stakeholders to the thing being evaluated (the evaluand), and that doing so is the best way to ensure both a high quality evaluation process and an accurate judgment about the quality of the evaluand.

EVALUATORS AS SERVANTS OF THE DEMOCRACY

When we think about social justice, our natural tendency is to think about things like why the poor have so little, or whether criminals receive their just

Visions of Quality, Volume 7, pages 181–200.

ISBN: 0-7623-0771-4

punishment. Often, we have strong opinions on these matters. Some people think that the poor will always be with us, so that it is hopeless to try to help them; but others think it is a moral imperative to better the condition of the poor. Some people think that we should lock up three-time offenders and throw away the key; but others think that more money should be spent rehabilitating them. In democracies such value positions are free to struggle and contend in the marketplace of ideas and at the ballot box (Jones, 1977). People are free to air strong opinions, and pluralism is respected. However, this diversity of values raises a dilemma for professional evaluators, one that has caused extensive debate in the field about such matters as: (a) How should values positions affect the evaluation? (b) Which values positions are allowed? and (c) How are diverse and sometimes conflicting value positions reconciled?

One proposed solution to these issues is that the evaluator should use his or her own judgment in adjudicating these many value disputes, and do so in a way that is explicit about the evaluator's own value commitments. This is described in many commendatory ways. For example, it is often said that such evaluators wear their values on their sleeves, being honest to the world about their value perspective. Doing so is taken to display understanding that science is not value free. Not doing so is taken to be something awful, like being a logical positivist who thinks that science and evaluation can be value free. Exercising this judgment based on one's personal values is taken to be a sign of courage and virtue – that if this is not done, and done explicitly, the evaluator is nothing but a lackey to the powers that be.

Our primary contention in this chapter is that such conceptions of the role of values in a socially just evaluation are fundamentally and profoundly misplaced. The reasons are both societal and technical:

(1) Democracies include diverse values positions. Relying primarily on evaluator judgment will result in censorship of some value positions. Censorship of such positions in the evaluation process ignores the legitimate roles of stakeholders, and it is antithetical to the democratic process.
(2) Subordinating stakeholder viewpoints to the personal judgment of an evaluator tends to disenfranchise and disempower stakeholders, especially those who are disadvantaged or oppressed.
(3) Personal values systems operate as perceptual filters that can bias professional evaluator judgments. Bias of all kinds is antithetical to good evaluation logic and methods. Inclusion of many values positions and of various stakeholder views offers the best insurance that bias is examined and identified.

Evaluators are not the moral police of the nation; and they can appear to be arrogant when they try to be its moral conscience. Conscience is a personal matter – it does not go by the board in the work we choose to do. However, we believe that evaluators best serve social justice when they act as servants of the democracy in which we live, and this chapter reminds readers of the simple ways that they can do this.

AN ASSUMPTION: PLURALISTIC EVALUATION FOR PLURALISTIC DEMOCRACY

Our fundamental contention is that we need pluralistic evaluation to serve a pluralistic democracy. So we must first understand how that democracy works. By most definitions, democracy is an ideological free-for-all (Jones, 1977) in which interest groups with different amounts of power, access, and resources contend both with and against each other to achieve preferred ends. Democracy *is* imperfect. Value positions *do* contend and strive in forums outside evaluation and will make *what they will* of the evaluation evidence; and free-ranging and even distorted uses of *all* information are the norm. Those with less access often *do* fail to have their will prevail. Take it or leave it, that is what democracy is. Evaluation can assist democracy when it provides one kind of discipline on this free-for-all by critically examining various value-laden positions that emerge during debates among positions. Socially just evaluation should not just serve a small subset of those value positions; rather it should aim to produce better information that all parties can use.

This commitment to pluralism increases our concern lest any legitimate stakeholder group – either disadvantaged or advantaged – be excluded from input into the information generated by an evaluation. Instead, evaluations require pluralistic input on values perspectives, just as they require pluralism in methods. We feel distinct discomfort when someone claims that social justice can be represented by just one viewpoint. No single stakeholder group has a monopoly on truth and justice, and any of them can oppress the rest. Too often, a group has claimed these attributes when ideology blinds them, or when a more cynical agenda is involved. Ideological extremists of many persuasions have claimed ownership of the single, true value position, or that they are the appointed ones whose judgment should hold sway. From the Holy Inquisition to the modern fascist and communist states, the results have been disastrous for social justice.

Pluralism has another justification as well, that it is the best defense against the tyranny of our own individual value biases. Personal values influence every aspect of the evaluation process, whether they are made explicit or not. No one familiar with the sociology of knowledge would disagree. However, each individual is often the least able to detect his or her own value biases. Indeed, ideologues often detect ideology only where it disagrees with their own preferred ideology, or even where it does not exist. Pluralism offers a better way to expose those personal value ideologies and biases, and it is achieved by listening to all the legitimate stakeholder groups. Those of us who recognize the substantial imperfections of our own beliefs and judgments are unwilling to create or synthesize values for others unless there is compelling experience to justify our doing so.

DESCRIPTIVE AND PRESCRIPTIVE VALUING

To be more technical about our argument, we differentiate between prescriptive versus descriptive theories of valuing in evaluation (Cook & Shadish, 1986, pp. 209–210; Shadish et al., 1991, pp. 47–53). *Prescriptive* theories defend the choice of one particular value (or set of values) as best. Normative ethics is a special case of prescriptive valuing pertaining to what is good for the human condition; but prescriptive theories of valuing are broader than ethics. An example of prescriptive valuing is saying that client needs should be the criteria by which we judge social programs (e.g. Scriven, 1995). Rawls' (1971) theory of justice is a prescriptive theory of ethics, which advocates that we ought to judge the ethics of something by the extent to which it favors "the interests of the disadvantaged" (House, 1995, p. 42). In both cases, a prescriptive approach to valuing in evaluation advocates one value as better than others, and uses that approach to valuing in evaluation.

Descriptive theory starts by describing the often-conflicting values held by stakeholders about the evaluand. It does not usually defend one of these values as best, and does not exclude values that disagree with a particular prescription. For example, a strictly Rawlsian approach would exclude the views of those who believe that citizens ought to be able to leave an estate to their children when they die, rather than having their estate distributed to the disadvantaged. Descriptive theories do not. The priority in descriptive valuing is on completeness and accuracy in describing those values. Those values should play the central (but not exclusive) role in creating value judgments about the evaluand. That role is *not*, and has never been in our view, simply to ask the stakeholders to do the evaluation themselves. Nor do we believe that evaluators should merely report stakeholder opinion about whether the evaluand is good. We agree with Scriven (1980, 1995) and House (1995) that the evaluator should take a more active role in drawing evaluative conclusions. We also believe, however, that *how* one goes about the process of assigning and synthesizing values is at least as important as the kind of values assigned. This is an area of importance that evaluators have neglected. A process that relies primarily on evaluator judgment will necessarily impose the evaluator's values at the expense of some stakeholders' values; a process that relies primarily on the major stakeholder prioritizations among values will be more democratic.

DESCRIPTIVE VALUING IN THE LOGIC OF EVALUATION

To further understand the role of descriptive values in evaluation, we must first recall what Scriven (e.g. 1980, 1994, 1995) has called the logic of evaluation. That logic advises us to: (a) select criteria of merit, (b) choose standards of performance for each criterion, (c) gather data, and (d) synthesize results into an evaluative statement. We claim that social justice is best facilitated when this logic relies more on descriptive than prescriptive values during the first, second, and fourth steps.

Suppose we are evaluating a long-term care program for the chronically mentally ill. Relevant stakeholders might include residents, their relatives, state and federal legislators, program managers, owners, staff, and neighbors. Each group holds certain values pertaining to long-term care, certain things (e.g. processes, outcomes) they think it would be good for long-term care to do. Figure 1 is adapted from Shadish et al. (1982) and illustrates how the stakeholders' values compare. The figure shows a multidimensional unfolding analysis in which smaller distances from stakeholder groups (signified by squares) to criteria of merit (signified by diamonds) indicates greater preference. Long-term care residents, for example, want a clean facility with nutritious food, an activity program, and staff who are interested in them – that is, good custodial care. These things are lower priorities for some other stakeholders, especially federal regulatory agency representatives. These latter stakeholders want patient care plans to be filed, extensive therapy programs, and efforts to move patients to independent community living – that is, cutting edge psychosocial treatment. The latter criteria are, however, lower priorities for the residents (Shadish et al., 1982).

The primary responsibility of a socially just evaluation is to make sure that each of these groups gets the information it needs about these matters to assess whether the program is doing what the group wants. To determine this, the evaluator must first take care to understand the values held by these stakeholders to the best of his or her ability – not always a straightforward task. The evaluator must then ensure that these values are represented in the various logic steps of evaluation. First, values must be represented among the criteria of merit to be used in the evaluation, for example, through selection of measures. Second, they must be represented in the standards for performance, for example,

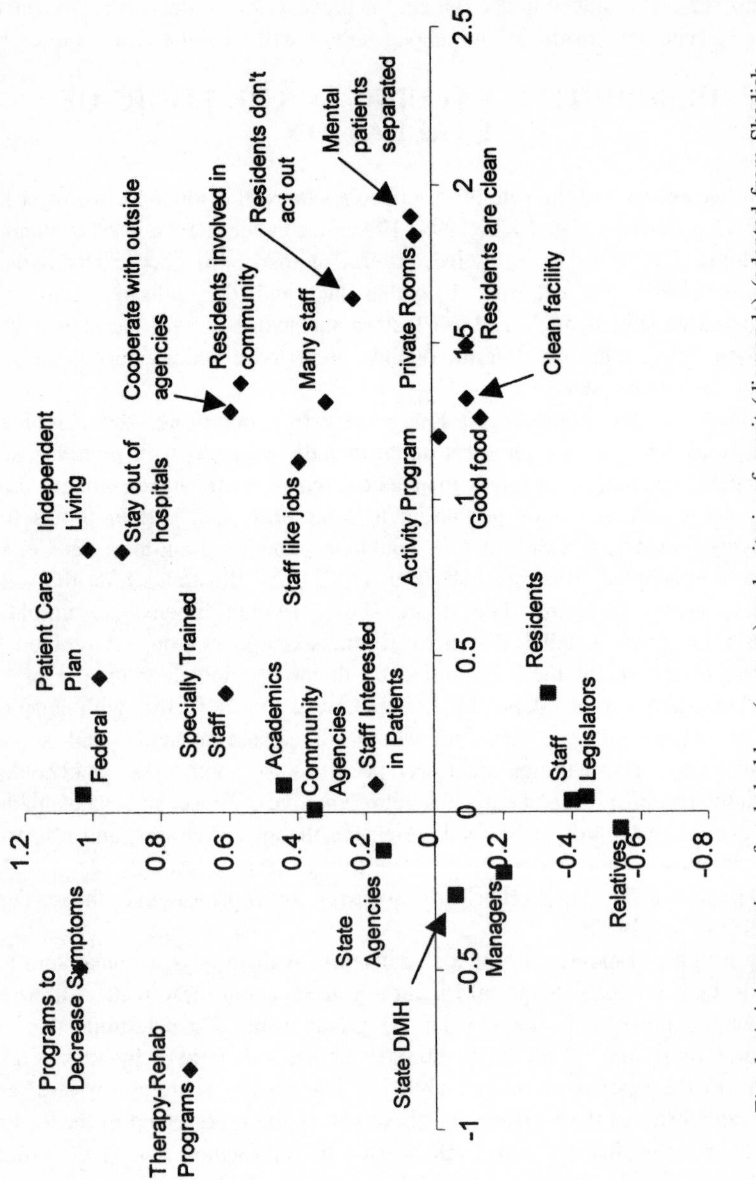

Fig. 1. A joint mapping of stakeholders (squares) and criteria of merit (diamonds) (adapted from Shadish, Thomas & Bootzin, 1982).

through selection of relevant comparison groups (e.g. by comparison to other mental health programs) or the setting of pertinent absolute minimum standards (e.g. minimum nutrition standards). Third, they must play a part in the synthesis of evaluative conclusions, for example, through multiple syntheses reflecting the major value profiles among stakeholders (e.g. long-term care facilities perform poorly by federal regulatory agency preferences but do better by patient preferences). That is what we mean when we advocate the use of descriptive values in evaluation.

Notice the crucial way in which this is not the same as just asking stakeholders their opinions – some stakeholder values may not be supported by the data gathered in step (c). This is the discipline that evaluation can provide to debates that emerge during the democratic free-for-all.

Prescriptive theories (e.g. Rawls, needs) would seem a logical way to generate socially just criteria of merit, standards of performance, and an evaluative synthesis. Yet this apparent relevance is deceiving. Many alternatives exist to Rawls' theory of justice, alternatives that would result in different criteria of merit and different conclusions about the merits of the program. No compelling empirical evidence supports these alternatives, nor is there any likelihood that it will emerge in the foreseeable future, because the pertinent data would require entire societies to adopt one or another prescription so we could empirically explore the consequences over many years. Lacking such evidence, the evaluator probably makes a choice among prescriptive theories based on personal judgment. This amounts to evaluation of the merits of programs based on personal beliefs about what is good for other people. The same argument holds for any other prescriptive approach, such as basing decisions on any particular evaluator's definition of needs. From such stuff springs scientific elitism.

By contrast under descriptive valuing, various stakeholder groups can have their values represented in the data gathering process, including those groups with whose values we might personally disagree. Descriptive valuing gives diverse stakeholders the information they need to fight their battles in the political and social arena. Evaluators are in the information business – we can give the body politic nothing more important than accurate information about the implications of diverse stakeholder values for programs and policies, even if they choose to use it in ways we would not approve. From such stuff springs a better functioning democratic pluralism. Inclusion of stakeholder values and concerns can, of course, be constrained both by the budget for data collection, and by our ability to identify the stakeholder groups. However, the principle of pluralism stands, and methods exist to accommodate these constraints, methods that prescriptive valuing would dismiss as irrelevant.

Like most who write about these matters, our advocacy of descriptive values is not absolute but rather one of relative emphasis. We reiterate that evaluation should rely *more on descriptive than prescriptive values*, and that descriptive values should *play the central (but not exclusive) role in creating value judgments*, a position we have taken ever since we first outlined this general argument (Cook & Shadish, 1986, p. 210; Shadish et al., 1991, p. 98). Because stakeholders' self-interest tends to promote narrow values positions, an exclusive reliance on descriptive values would be bad if values that ought to be present in the debate are not represented among any of the stakeholders. To help remedy such problems, evaluators must keep sight of the *variety* of prescriptive frameworks, to provide another source of criticism of the array of value perspectives available among stakeholders. But none of those prescriptive frameworks should be imposed at the expense of the primacy of descriptive stakeholder values.

WHO SPEAKS ABOUT VALUES?

Over-reliance on prescriptive valuing can be actively harmful and undemocratic because it becomes too easy to ignore stakeholders' own statements of their value positions. Stakeholders, including disadvantaged stakeholders, often articulate their own needs and value positions very well without the judgment of evaluators as a mediator. Moreover, they often have important information to offer about specifics. For such reasons, prescriptive theories cannot supplant the need to consult stakeholder groups. Rawls (1971), for example, generally tells us to distribute goods and resources to the benefit of the disadvantaged; yet would not *some* of the disadvantaged have said this themselves, and perhaps told us much more specific information in this regard than Rawls' theory could hope to provide? And if the evaluator does not consult the stakeholders, how can we understand their preferred ways and means to handle the distribution? The devil is in the details, which a simple assessment of values would ignore, but which is critical to the success of resulting policy and program change. Even when distributive justice has been the direct goal of social programming, a failure to consult disadvantaged groups directly has had disastrous consequences, for example in international development programming and in environmental health (Leviton et al., 1997).

Needs-based prescriptions have similar limitations. Scriven tells us to generate criteria of merit from a needs assessment, but would not *some* of the stakeholders be able to state their own needs? Why should the evaluator do this for them, and who defines their "real" needs? The chronically mentally ill in Shadish et al. (1982) were reasonably articulate about the value they placed on criteria such as a clean, stimulating place to live with nutritious food. And sometimes, formal needs assessments set priorities that are in stark contrast to those of communities. This occurs in public health for example, where data on disability and death can dictate priorities quite different from the ones communities strongly prefer (Leviton et al., 1997). Should evaluators claim to speak for these stakeholder groups?

Prescriptive valuing methods tend to supplant stakeholders' own statements about what they value. Community organizers know this, and consequently, some have come to view formal needs assessment as part of a process in which professionals disempower communities and reduce their ability to articulate their own needs and wants (McKnight, 1995). Such concerns led Saul Alinsky (1989) to be profoundly skeptical of the corrupting influence of those who would impose their own judgments:

> The Have-Nots have a limited faith in the worth of their own judgments. They still look to the judgments of the Haves. . . . The Have-Nots will believe them where they would be hesitant and uncertain about their own judgments. . . . It is a sad fact of life that power and fear are the fountainheads of faith (p. 99).

Prescriptive values that oppress the disadvantaged led Paulo Freire (1970) to develop more sensitive methods to engage oppressed groups in a respectful way and to elicit from them an improved understanding of their social problems. Some pragmatism is also in order. Low-income communities are often sick to death of professional prescriptions and are correctly vociferous about establishing their own priorities for social action and for evaluation (Leviton, 1994; Leviton et al., 1997; Maciak et al., 1998). An evaluator who tried to determine needs and wants from his armchair might find himself riding out of town on something not quite so comfortable.

BIAS IN PERSONAL VALUES

When we think about social justice in evaluation, we inevitably bring our personal opinions and understandings to that job. Indeed, we have seen novice evaluators who regularly become indignant over flaws they perceive in programs. It is natural to feel such indignation, especially when it affects the disadvantaged. Yet the evaluator must be vigilant lest such personal understandings dominate the conclusions about value that we draw from an evaluation. Relying too much on such individual understandings, in values positions as in anything else, is antithetical to the idea of evaluation as disciplined inquiry. Disciplined inquiry helps to counteract biases of many kinds, whether we are speaking of wishful thinking and rhetoric about program utility, or perceptual and methods biases that affect causal attributions about program effects.

Disciplined evaluative inquiry requires constant vigilance in identifying the personal perceptual filters we use in evaluation. We do not perceive the values positions of others directly, just as we do not measure "poverty" directly, and just as we infer, not prove, a causal relationship. Our personal values constitute a perceptual filter through which we interpret the values of others. For decades, it has been understood that values are socially constructed (albeit with reference to data), not objectively "there". The evaluator does not share the stakeholders' social reality and may not even understand it well. For this reason, ethnography takes time. For this reason, Freire (1970) had to commit time to develop a shared social reality with oppressed slum dwellers. Prescriptive valuing is antithetical to this process. When priority is given to prescriptive valuing over descriptive valuing, the perceptual filters of the evaluators remain unexamined; this can do violence to the real values positions of the various parties. Stakeholder values can sometimes be subtle, as seen in the careful work of Olsen et al. (1985) concerning development on the island of Molokai. By seeking input from the variety of stakeholders and value positions, assumptions about values can be actively tested.

SINGLE VERSUS MULTIPLE DESCRIPTIVE SYNTHESES

Our position on descriptive valuing seems to have generated the most disagreement regarding the fourth step of Scriven's logic-especially our recommendation (e.g. Shadish et al., 1991, pp. 100–101) for multiple syntheses based on stakeholder-generated value profiles. To repeat a few minor points that we have stated before, we agree that single syntheses will sometimes be *desirable*, especially when strong value consensus exists among all stakeholders, and when all the important value dimensions are reflected in that consensus. We also agree that single syntheses are always *possible*; but to generate one when differences in value prioritization exist will often require suppressing the values of one or more stakeholder groups, which we reject as socially unjust. Returning to the example of long-term care for the chronically mentally ill, we might conclude that long-term care is good if one values the provision of acceptable levels of custodial care at a reduced cost from the mental hospital. In other words, it is good from the perspective of the patient, as well as those whose views are similar to the patient such as relatives, staff, and local legislators. However, long term care is not good to the extent one values reducing symptomatology or returning the resident to independent community functioning. In other words, it is not so good from the perspective of many federal agencies. It is difficult to see how forcing a single synthesis on this case does justice to the diverse values held by relevant stakeholders, given what we know about such care (e.g. Bootzin et al., 1989; Shadish, 1986; Shadish & Bootzin, 1984). After all, what would such a synthesis look like? If we say long-term care is good, then the federal stakeholders will object that it is not from their perspective (and they would be right). If we say it is bad, the patients would object that it is good from their perspective (and they would be right, too). If we say it is both good and bad in different ways, that is just putting two syntheses in one sentence, which doesn't really solve the underlying problem of value synthesis.

For such reasons, the best practice in policy analysis and in business management is usually to avoid the single value synthesis. Some evaluators argue that to produce a single synthesis from multiple value systems, the prescriptive valuing approach might not ignore stakeholders, but instead strive for consensus. Yet even this is not advisable as a general strategy. A false or premature consensus on values has long been known to be harmful; eliciting real differences in assumptions and values can have a healthy long term payoff for organizations and for society (Mitroff & Kilman, 1978). Seeking a single synthesis, even when one sets priority on the needs of the disadvantaged, may be dangerous when there are important factions within the apparent single stakeholder group (e.g. Gray, 1989). These differences can go unrecognized, appear regularly, and can even promote social injustice, unless care is taken to do good descriptive valuing.

APPLICATIONS AND EXAMPLES ABOUT SYNTHESIS

It helps to illustrate our arguments about multiple syntheses to respond in detail to some applications proposed by House (1995), who has taken particular and direct objection to our arguments in favor of multiple syntheses. He claims that the evaluator usually should reach a single synthesis about program worth, and that the synthesis should be generated using the evaluator's contextual judgment. He offers several examples of single syntheses being done, trying to make the point that conflicting values do not impede a single synthesis; in our view, however, all the examples he cites contain fatal flaws. Two of the examples he uses are the fact that: (a) the members of university tenure committees routinely reach a single evaluative decision about tenure, and (b) juries in law courts routinely reach a single verdict representing their evaluation of the evidence. Unfortunately, these are not examples of professional evaluators reaching single syntheses. The differences in social roles between: (a) evaluators, versus (b) juries and faculty tenure committees are crucial. Society has specifically instructed faculty members in the first example, and juries in the second example, to come up with one and only one evaluative judgment. That is their job, and we view it as a failure if they do not do so. We even have a label for the failure in the case of a hung jury. By stark contrast, it is absolutely crucial to see that society has *not* empowered professional evaluators to reach such judgments; and society would not view it as a failure if we reach multiple syntheses. In fact, society has *never* even deliberately sought out and hired a professional evaluator (*qua* evaluator) to make a tenure decision or decide a legal issue. The fact that evaluators may have participated in such work is just a coincidence of social roles (some faculty members happen to be professional evaluators, as do some jurors), and it is improper to cite that participation as evidence for the single synthesis approach in professional evaluation (it *is* evidence for single syntheses in tenure and jury decisions, of course, but that's not the point). What we need is evidence that society has asked evaluators to reach only one synthesis, specific evidence like law, regulation, or instruction from a duly constituted legal body, on a widespread basis. The evidence cannot simply be an evaluator's opinion.

Moreover, close inspection of the faculty evaluation example suggests that the single synthesis of diverse values that House claims this example supports is actually the antithesis of taking multiple stakeholder values into account. House (1995) said: "In the early 1980s the University of Illinois was strongly research oriented. Although the criteria were research, teaching, and service (the same as elsewhere), the only thing that counted was research" (p. 35). The best lesson one can draw from this is that if only one criterion counts, then reaching a single synthesis is easy. However, House clearly says this choice to use only one criterion was reached by using only the priorities of the "University Board of Trustees" (p. 34) in this process. What happened to stakeholders other than the University Board of Trustees? What happened to students and parents who think teaching counts? What happened to legislators who think faculty do not spend enough time in the classroom? This example confuses the interests of a single client (the board of trustees) with a single synthesis of the many stakeholders to the tenure process. Taking all such stakeholders seriously would make it much more difficult – and probably impossible in many cases – to reach a single synthesis.

All the other examples that House (1995) offers of single syntheses reveal equally fatal problems. One example is a conflict in an evaluation of health care services for the elderly in Texas where some stakeholders said the elderly lacked knowledge of the services but the elderly said the services were not accessible. House proposes that to obtain a synthesis "the evaluators should try to determine which group was correct about the services" (p. 43). But this remark misses the difference between *facts* and *values* that is the heart of the debate. Whether the elderly knew about services, and whether the services were available, are matters of fact that the evaluator can indeed study productively and usually resolve with data. However, this example does *not* present a *value* conflict (as opposed to a factual conflict) to resolve; all parties seem to agree that services should be accessible and the elderly should know about them. The best lesson we can draw from this example is that in the atypical case where there is no disagreement about values, synthesis is much easier.

In his only really conflictual example, House (1995) describes a hypothetical evaluator who is faced with three interest groups that value equality, excellence, or entitlement, respectively. This is similar to a classic value tradeoff between liberty and equality; it being difficult to maximize both simultaneously because enforcing equality tends to impinge on liberty, and allowing liberty tends to reduce equality. House claims that "such value issues can be resolved rationally" (p. 45), but does not say how, only advising that "such a synthesis requires data collection and complex deliberation" (p. 45). This is inadequate. Simply telling the evaluator to gather data, deliberate, and then use judgment to synthesize results into one position does not solve the problem of how to do this without trampling over the interests of one or more groups.

Finally, House (1995) offers the following example to demonstrate that multiple syntheses are wrong. He says:

> Let us push the limits of the argument with an extreme example. . . . What if the evaluation determined that the medical services provided in Texas contained strong racist elements? Surely, that aspect of the program would be condemned by the evaluators. Evaluators wouldn't say, "The program is good if you are a racist, but bad if you are not", which would be making different syntheses for different groups with different values (p. 44).

House calls this an extreme example of the synthesis issue, but it is not for several reasons. First, House did not tell us how he would reach a single synthesis here, or what that synthesis would be. For example, he did not say the entire program was bad because of the racist element. If he did, then for those of us that believe most U.S. institutions contain nontrivial racist elements, there is little point in further evaluation. We can simply conclude that those programs are all bad, and be done with it. In fact, House did the opposite. He asked us to condemn "that aspect of the program", not the whole program; and he would presumably have a different conclusion about other aspects of the program. That is perfectly consistent with multiple syntheses.

Second, House describes a value conflict between stakeholders about racism here (two stakeholders, one value), but ignores conflicts that involve prioritization between racism and other values (multiple stakeholders, multiple values). These latter conflicts are the most common, and they make the value synthesis task far more difficult even in the extreme example he claims to solve. Suppose everyone agrees that this racist program had done more to improve the health of Hispanic populations than any other program in state history? Is it still simply "bad" because it contained racist components, or would it be better to reach the multiple syntheses that: (a) the racist components of this program are bad, but (b) the accomplishments for the health of the Hispanic population are good?

Third, this example cannot serve as a clean test of single versus multiple syntheses because it confounds the synthesis issue with the evaluator's duty as both a citizen and as an evaluator to uphold the law. Certain forms of racism are illegal, which the evaluator cannot ignore any more than one could ignore the discovery of embezzlement, child abuse, or other illegal activities – even if the program was perfect on all other counts. In Scriven's terms, if it is an illegal activity of a substantial nature, it may require concluding that the program fails the evaluation because it fails to meet a pertinent absolute standard for minimum acceptable performance on some criterion. But most evaluations do not involve the discovery of law breaking, and they are harder to resolve.

Fourth, if the racist element is not illegal (e.g. Anglo providers prefer to serve Anglo clients, and Hispanics prefer to serve Hispanics, but all clients get served, and all parties think the situation is as good as one is going to get in the context), then it is not the evaluator's business to issue a blanket negative evaluation of the program because the evaluator does not like the racism that is implicit in this pattern. Of course, the evaluator should describe this racism (and of course that will be done in negative terms, because racism is a negative feature), along with all the other good and bad that was found, but basing the whole evaluation synthesis on such a racist element serves only to make the evaluator feel self-righteous. Finally, House admits this is an extreme example. But extreme examples are atypical by definition, so even if none of the other objections to this example held, it would still tell us very little about what evaluators ought to do in more typical cases of value conflict.

Perhaps, though, we are not as far apart from House as these examples might lead one to think. For example, House (1995) says at one point that "the evaluator can usually arrive at legitimate single (if highly qualified) judgments" (p. 41). A few pages later he returns to the example of Sesame Street that we have long used to illustrate multiple value conclusions (e.g. Shadish et al., 1991, p. 101); and he acknowledges Sesame Street might be judged good from some perspectives because it raises all reading scores, but bad from other perspectives because it increases the gap between advantaged and disadvantaged children's reading scores. He concludes, "one might arrive at a synthesis judgment that the program is good, but not for decreasing the gap between groups" (p. 43). We would call that multiple syntheses (two conclusions); he would call it highly qualified. What it is not is a single synthesis, that Sesame Street is either good or bad, but not both. Otherwise, House's single synthesis becomes trivial, reducing to nothing more than including all positions in a single sentence.

Scriven's position about multiple versus single syntheses is less clear, but we read his recent work as more compatible with multiple syntheses reflecting multiple stakeholder interests. For example, he frequently shows how syntheses are reached in automobile (or other product) evaluations. But close inspection shows these examples are usually multiple syntheses. In speaking of the role of acceleration, Scriven (1994) says that such acceleration "is a merit in a car" (p. 372), but also "if you like, you can qualify the claim to refer only to cars for people interested in performance" (p. 372), which is a stakeholder-specific synthesis. He also acknowledges the legitimacy of dividing cars into "classes" based largely on price and features (e.g. trucks, sports vehicles, etc.). But those classes always reflect differences in stakeholder values and interests (e.g. some of us can afford to spend lots of money, and others cannot). Those differences could have been included as criteria of merit (e.g. a less expensive car is better, all things being equal) instead of using them to create car classes. However, doing so would raise a value conflict among the criteria, for example, it being difficult to maximize performance (or a whole set of other automobile characteristics) while minimizing price. Dividing cars into classes, then, is really no different from dividing car stakeholders into classes based on their interests and values (e.g. those who want trucks, or those who can afford luxury cars). It is multiple synthesis that is perfectly consistent with the descriptive approach we advocate.

EVALUATION IN A PLURALISTIC INTEREST GROUP DEMOCRACY

To return to our main theme, it is a tenet of democracy that it works best when all citizens have free access to the information they need to make decisions. Justice is about the fair distribution of resources. If information is an essential resource, then it can rarely be socially just in a democracy to suppress information about some value perspectives. To refuse to include stakeholders' values as criteria of merit, or to suppress those values during the final evaluative synthesis, necessarily deprives those stakeholders of information they need. In such a context, descriptive valuing is better suited than prescriptive valuing to maximizing the evaluative information that all citizens need to participate in democracy. Evaluators who suppress values, even in the name of prescriptive valuing, are fostering socially unjust evaluations.

House (1995) says, "I believe the professional evaluator is authorized to make such judgments as part of the role" (p. 43). But he presents no evidence to support that belief; and we have argued that no such evidence exists. Unlike the case for juries or for faculty tenure committees – where enormous legal and regulatory documentation is available to justify the single syntheses made in those cases – no law, regulation, contract, or any other authoritative source ever gave evaluators an explicit instruction to come up with a single synthesis even once, much less usually. This omission may well be intentional, because citizens in a pluralistic democracy may not want the evaluator to reach just one synthesis at the expense of overlooking important stakeholder objections. Nor can evaluators fairly claim that some philosophical logic of evaluation gives them the right to create one synthesis because nothing in that logic requires a single synthesis – and because evaluators invented that logic, giving the appearance if not the fact of a self-serving argument. Nor can they simply use their own beliefs that they have the right to make such single syntheses. Professional evaluators, by virtue of being professionals, must abide by the legal and regulatory strictures that society has set for their functioning.

One might return, then, to the question of what role is left for the professional evaluator? Are evaluators reduced to simply reporting stakeholder evaluations? Not at all; that is merely a caricature used by opponents of descriptive valuing. Evaluators have plenty of tasks to do that have been legitimated through law, regulation, or contractual request. First, the third step of the logic of evaluation, gathering data, is entirely the purview of the professional evaluator who specializes in methods for this task, and for evaluating the knowledge generated by the third step. This is the "fact" part of the fact-value distinction, and few evaluators dispute that evaluators have special skills in identifying those "facts" (e.g. whether the elderly knew of the medical services in House's Texas example). There is ample evidence that society has asked evaluators to do this in regulations, laws, or contracts. Second, the methods that are required for doing the other three steps of the logic of evaluation (e.g. a stakeholder analyses) almost always require a professional evaluator who specializes in those tasks, as well. As long as these methods are implemented in a way that does not deliberately suppress some values, they are entirely appropriate. Third, the professional evaluator is usually the one who knows how to interpret the technical aspects of the evaluation, such as whether the available evidence supports a causal inference that the treatment caused the outcome, or whether the sampling methods allow a confident generalization to a population of interest. After all, most evaluators have years of socially certified training in such tasks through graduate school coursework or continuing education. Society cedes us this expertise (although sometimes we disagree among ourselves so much that society throws up its hands in disgust!). It is perfectly reasonable, perfectly consistent with our view of social justice, for evaluators to usurp these tasks, because society has asked us to do so.

CHALLENGING DEMOCRATIC PLURALISM

Perhaps the most compelling criticism of the descriptive approach that we advocate in this chapter is to challenge the moral force of democratic pluralism, for that would undercut any claim that it is moral to serve democratic pluralism. The challenge has two forms. One is to note that a particular pluralistic interest group democracy is far from perfect, especially because some stakeholders have more power than others to act on the evaluative information we give them. This is true. However, it is also true of every other form of government currently practiced. Imperfections do not require rejecting the system. Indeed, to the contrary, Saul Alinsky (1989) described community organizers in terms that ring equally true for evaluators:

> As an organizer I start from where the world is, as it is, not as I would like it to be. That we accept the world as it is does not in any sense weaken our desire to change it into what we believe it should be – it is necessary to begin where the world is if we are going to change it to what we think it should be. That means working in the system (p. XIX).

To think otherwise is to argue that evaluators should not serve whatever political system in which they live or work, democratic or not, even if they believe that political system to be the best available, however flawed. We think this objection will usually fail because it requires the critic to make one of two usually untenable arguments, either that society wants professional evaluators to serve only an extrasystemic criticism role, or that the moral offenses of the political system are so egregious that it is fundamentally wrong to support it in any way. The first of these arguments is, as we pointed out earlier, unsupported by reference to any law, regulation, contract, or other socially-legitimated source, with one possible exception. That exception is the academic evaluator. The doctrine of academic freedom is both institutionalized and aimed at allowing even that speech that is patently critical of the political system. But this is clearly a limited claim at best, not applicable to most of professional evaluation, and even then probably derivative of the history of academe rather than essential to the profession of evaluation. The second of these arguments will always be a judgment call that each evaluator will have to make as a citizen, for instance, refusing to serve certain types of political systems (e.g. repressive dictatorships) or refusing to support certain aspects of an otherwise acceptable political system (e.g. as in the civil rights era in the U.S.). But no matter how that judgment call is made, it is part of being a citizen, which all professional evaluators happen to be.

The second challenge to democratic pluralism might come from those parts of the evaluation community that live or work in countries with political systems that are patently not democratically pluralistic. As U.S. evaluators, is it ethnocentric of us to assume that democratic pluralism is the best of the currently available forms of government? For example, the "Guiding Principles for Evaluators" (Shadish et al., 1995) were developed by U.S. evaluators with experience primarily in U.S. political contexts. Consequently, that document contains explicit references to serving democracy (e.g. Section III.E.3. on the importance of freedom of information in a democracy) and implicit assumptions about serving democratic pluralism (e.g. Section III.E.1. on considering a range of stakeholder values). We did not fully appreciate this until after the final product was accepted by vote of the AEA membership. In retrospect, however, it raises the issue of why such assumptions should hold, and what guiding principles should apply to evaluators whose political system is radically different, for instance, a dictatorship of either the left or the right where freedom of speech is more severely curtailed. Because the present authors are so deeply immersed in our political system, we are not in a good position to take up this challenge, and would like to hear from international colleagues about it. But if such a challenge were successfully mounted, it might constitute a very interesting challenge to our view of the moral justification for descriptive valuing.

SUMMARY

One paragraph in House (1995) probably best captures what we suspect are the misunderstandings and the real differences between his position and ours. In responding to our previous writings, he says of us:

> Believing that evaluators cannot or should not weight and balance interests within a given context . . . suggests that human choice is based on non-rational or irrational preferences, not rational values, and that these things cannot be adjudicated by rational means. It implies a view of democracy in which citizens have irrational preferences and self-interests that can be resolved only by voting and not by rational means such as discussion and debate. A cognitive version of democracy, one in which values and interests can be discussed rationally and in which evaluation plays an important role by clarifying consequences of social programs and policies, is much better (p. 46).

If we were to rephrase this paragraph to better fit our own beliefs, we would say:

> Evaluators can weight and balance interests, and when value conflicts are minimal, they should do so; otherwise they should not, if this means suppressing some of those interests. Human choice is based on non-rational, irrational, and rational grounds, usually all at once. While some of these choices can be adjudicated by rational means, the irrational preferences are usually far more pervasive and far less tractable than the rational ones, so successful rational adjudication, particularly by scientists, is the exception rather than the rule. In a democracy, citizens really do have irrational preferences and self-interests. Lots of

> them; and they are entitled to them. Sometimes they can be resolved rationally through discussion and debate, and when that is possible, it should be done. But more often than not, they can only be resolved by voting or some other method that voters legitimate. Evaluation can play an important role by clarifying consequences of social programs and policies, especially concerning outcomes that are of interest to stakeholders. But this is unlikely to lead to a so-called cognitive democracy in which values and interests are adjudicated rationally.

When it comes to values, we think the world is a much less rational place than House seems to think. We agree with him that it is good to try to increase that rationality, within limits. When stakeholders differ about important values, it can sometimes make sense for the evaluator to engage them to see if they can find some common ground (although that task frequently extends well beyond the evaluator's training and role). When the needs of the disadvantaged look to be omitted from an evaluation, by all means the evaluator should make note of this and consider a remedy (descriptive valuing usually is such a remedy). But all this is within limits. If values were entirely rational, they would not be values by definition. So there is a limit to what the evaluator can hope to accomplish, and we suspect that limit is far lower than House's admirably high aspirations. Trying to impose a cognitive solution on value preferences that have a substantial irrational component is the problem democracy was originally invented to solve. To paraphrase House's (1995) last sentence: Disaster awaits a democratic society that tries to impose such a solution.

REFERENCES

Alinsky, S. D. (1989). *Rules for radicals: A practical primer for realistic radicals*. New York: Vintage Books.

Bootzin, R. R., Shadish, W. R., & McSweeny, A. J. (1989). Longitudinal outcomes of nursing home care for severely mentally ill patients. *Journal of Social Issues*, *45*, 31–48.

Cook, T. D., & Shadish, W. R. (1986). Program evaluation: The worldly science. *Annual Review of Psychology*, *37*, 193–232.

Freire, P. (1970). *Pedagogy of the oppressed*. New York: Herder and Herder.

Gray, B. (1989). *Collaborating: Finding common ground for multiparty problems*. San Francisco: Jossey-Bass.

House, E. R. (1995). Putting things together coherently: Logic and justice. In: D. M. Fournier (Ed.), *Reasoning in Evaluation: Inferential Links and Leaps* (pp. 33–48). San Francisco: Jossey-Bass.

Jones, C. O. (1977). *An introduction to the study of public policy* (2nd ed.). North Scituate, MA: Duxbury Press.

Leviton, L. C. (1994). Program theory and evaluation theory in community-based programs. *Evaluation Practice*, *15*, 89–92.

Leviton, L. C., Needleman, C. E., & Shapiro, M. (1997). *Confronting public health risks: A decision maker's guide*. Thousand Oaks, CA: Sage Publications.

Maciak, B., Moore, M., Leviton, L. C., & Guinan, M. (1998). Preventing Halloween arson in an urban setting. *Health Education and Behavior*, *25*, 194–211.

McKnight, J. (1996). *The Careless Society: Community and Its Counterfeits*. New York: Basic Books.

Mitroff, I. I., & Kilman, R. H. (1978). *Methodological approaches to social science*. San Francisco: Jossey-Bass.

Olsen, M. E., Canaan, P., & Hennessy, M. (1985). A value-based community assessment process: Integrating quality of life and social impact studies. *Sociological Methods and Research*, *13*, 325–361.

Rawls, J. (1971). *A theory of justice*. Cambridge, MA: Harvard University Press.

Scriven, M. (1980). *The logic of evaluation*. Inverness, CA: Edgepress.

Scriven, M. (1994). The final synthesis. Evaluation *Practice*, *15*, 367–382.

Scriven, M. (1995). The logic of evaluation and evaluation practice. In: D. M. Fournier (Ed.), *Reasoning in Evaluation: Inferential Links and Leaps* (pp. 49–70). San Francisco: Jossey-Bass.

Shadish, W. R., & Bootzin, R. R. (1984). Nursing homes: The new total institution in mental health policy. *International Journal of Partial Hospitalization*, *2*, 251–262.

Shadish, W. R. (1986). Planned critical multiplism: Some elaborations. *Behavioral Assessment*, *8*, 75–103.

Shadish, W. R., Cook, T. D., & Leviton, L. C. (1991). *Foundations of program evaluation: Theories of practice*. Newbury Park, CA: Sage Publications.

Shadish, W. R., Newman, D. L., Scheirer, M. A., & Wye, C. (Eds) (1995). *Guiding Principles for Evaluators*. San Francisco: Jossey-Bass.

Shadish, W. R., Thomas, S., & Bootzin, R. R. (1982). Criteria for success in deinstitutionalization: Perceptions of nursing homes by different interest groups. *American Journal of Community Psychology*, *10*, 553–566.

13. DEFINING, IMPROVING, AND COMMUNICATING PROGRAM QUALITY[1]

Joseph S. Wholey

ABSTRACT

Throughout the world, both in government and in the not-for-profit sector, policymakers and managers are grappling with closely-related problems that include highly politicized environments, demanding constituencies, public expectations for high quality services, aggressive media scrutiny, and tight resource constraints. One potential solution that is getting increasing attention is performance-based management or managing for results: the purposeful use of resources and information to achieve and demonstrate measurable progress toward agency and program goals, especially goals related to service quality and outcomes (see Hatry, 1990; S. Rep. No. 103-58, 1993; Organization for Economic Cooperation and Development, 1996; United Way of America, 1996).

INTRODUCTION

This paper illustrates the application of performance-based management approaches to the problems of measuring and evaluating program quality, improving program quality, and communicating the value of program activities in ways that lead to improved programs and support policy decision making.

Visions of Quality, Volume 7, pages 201–216.

ISBN: 0-7623-0771-4

We define a program as a set of resources and activities with one or more common goals. Thus "program quality" may refer to the quality of any project, function, policy, agency, or other entity that has an identifiable purpose or set of objectives.

As is true of other papers in this volume, this paper takes the position that program quality is never an objective reality out there waiting to be discovered, measured, and evaluated. Instead, program quality is socially constructed reality (Berger & Luckmann, 1966). Within specific contexts, the meaning of program quality is maintained (and modified) through day-to-day experience. Performance-based management seeks to define and refine the meaning of program quality through strategic planning and performance planning – and to reinforce the meaning of the term through ongoing performance measurement and performance reporting.

Important dimensions of program quality may be found at one time or another in program inputs, program activities, program outputs, or intended or unintended outcomes – both in programs that are the responsibility of a single agency and in *crosscutting programs* in which two or more agencies share common goals. In this day and age, program quality increasingly refers to the extent to which programs achieve progress toward outcome-related goals. The meaning of program quality may be redefined as a result of day-to-day experience, planning, or program evaluation, any of which may reveal other performance dimensions that should be incorporated.

The paper examines three topics that are central to performance-based management: developing a reasonable level of agreement on what constitutes program performance; measuring and evaluating program performance; and using performance information to improve program performance, provide accountability, and support policy decision making. It explores how policymakers and managers can use planning, performance measurement, and program evaluation in defining program quality, improving program quality, and communicating the quality of programs to those within the program and to external stakeholders affected by or interested in the program. Examples are drawn from the teaching and evaluation professions and from efforts to implement the Government Performance and Results Act of 1993 (Results Act), which is central to current performance-based management efforts at the federal level in this country.

QUALITY MANAGEMENT

Though *total* quality management appears impractical and the term TQM is less in vogue these days, quality management is alive and well, reincarnated within current initiatives variously labeled as *performance-based management*, *managing for results*, or *focusing on outcomes* (see U.S. General Accounting Office, 1996; United Way of America, 1996). The key steps in quality management are: (1) developing a reasonable level of agreement on the meaning of program quality; (2) developing performance measurement systems and program evaluation studies that provide information that is sufficiently timely, complete, accurate, and consistent to document performance and support decision making; and (3) using information on program quality to improve program quality, communicate the value of program activities, and support resource allocation and other policy decision making. This section examines each step in turn.

Planning: Defining Program Quality

In any specific context, the term program quality may have many meanings. Though program quality is in the eye of the individual beholder, quality management requires some shared understanding of the term. Priorities must be established (Timpane, 1998). If the meaning of program quality is too diffuse, the term will mean little or nothing to program managers and those engaged in service delivery.

Program quality may focus on *inputs*; for example, the quality of the teachers who assemble each school day or the quality of the evaluators who staff a particular unit or a specific study. Program quality may focus on *activities* or *processes* for converting inputs into outputs and then into outcomes; for example, the extent to which the teaching process complies with applicable guidelines or the extent to which the evaluation process meets Joint Committee standards relating to utility, feasibility, propriety, and accuracy (Joint Committee on Standards for Educational Evaluation, 1994). Program quality may focus on outputs; for example, classes taught or evaluation studies completed. In current performance-based management initiatives, program quality focuses especially on *outcomes*. Program quality may focus on *intermediate outcomes*; for example, the extent to which schoolchildren learn to read prose or write poetry, client satisfaction, or the extent to which agencies implement evaluation recommendations. Program quality may focus on *end outcomes*; for example, students' entry into college or a profession, or the extent to which programs improve or policies change after evaluations are completed.

The meaning of program quality may change as day-to-day experience, planning efforts, benchmarking comparisons, basic or applied research, evaluation studies, or experimental trials bring other performance dimensions into focus (see Weiss & Morrill, 1998). To meet concerns raised by Radin (1997) and others, the notion of program quality should be extended to include important *unintended outcomes*; for example, teaching to the test, failure to provide fair treatment, costs incurred as school districts respond to program requirements, or costs incurred by agencies as they implement evaluation recommendations. Agencies can bring important unintended outcomes into a quality management framework by getting a reasonable level of agreement on goals and strategies for minimizing or controlling those unintended outcomes.

Especially in policy decision making, program quality may focus on *net impact*: what difference a program has made. Net impact can be measured through program evaluation studies that compare program outcomes with estimates of the outcomes that would have occurred in the absence of the program.

To implement useful management systems focused on program quality, agencies must first achieve a reasonable level of agreement among senior officials, managers and staff, and other key stakeholders on agency or program goals and on the inputs, activities, and processes that will be used to achieve the goals. Public and not-for-profit agencies can often achieve such agreement through strategic planning processes that include broad consultation with managers and staff, those who influence the allocation of needed resources, those who make other policy decisions affecting the agency or program, and other key stakeholders affected by or interested in agency or program activities. Agencies should define program quality broadly enough to cover the key performance dimensions – the inputs, activities, outputs, intermediate outcomes, end outcomes, and unintended outcomes – that are of importance to key stakeholders. Agencies can use logic models and the evaluability assessment process to help ensure that they have identified relevant elements of the program design: key program inputs, activities, outputs, intermediate outcomes, end outcomes, important unintended outcomes, assumed causal linkages, and key external factors that could significantly affect performance (see Wholey, 1994).

Examples illustrating the importance (and difficulty) of defining, measuring, and evaluating program performance can be found in the current efforts of federal agencies. In the Results Act, inspired by the growing use of performance measurement at state and local levels and throughout the world, Congress prescribed consultation and planning to identify agency and program goals and annual reporting on performance. Contrary to efforts 30 years earlier under the planning-programming-budgeting system, current federal-level performance-based management efforts are grounded in statute and are focused on agencies and programs that are the responsibility of senior managers and specific congressional committees. The required planning and performance measurements are intended to improve internal management; help managers improve service delivery; improve program effectiveness and public accountability by promoting a focus on results, service quality, and customer satisfaction; improve policy decision making; and improve public confidence in government (S. Rep. No.

103-58, 1993). With minor exceptions, federal agencies' strategic plans, annual performance plans, and annual performance reports are public documents.

Beginning in September 1997, the Results Act required federal agencies to submit strategic plans. The strategic plans were to: (a) contain mission statements covering the major functions and operations of the agency; (b) identify strategic goals and objectives, including outcome-related goals and objectives, for the major functions and operations of the agency; (c) describe how the goals and objectives are to be achieved, including a description of the resources and processes required to meet the goals and objectives; (d) describe how the performance goals included in the agency's annual performance plans would be related to the agency's strategic goals and objectives; (e) identify the key external factors that could significantly affect the achievement of the goals and objectives; and (f) describe the program evaluations used in establishing or revising strategic goals and objectives, with a schedule for future program evaluations. In developing their strategic plans, agencies were required to consult with Congress and solicit and consider the views of other key stakeholders affected by or interested in agency activities. Strategic plans are to cover at least six years: the fiscal year it is submitted, and at least five years forward. The strategic plans are to be updated as often as is appropriate – at least every three years.

With the submission of each proposed budget beginning with the budget for fiscal year 1999, the Results Act required submission of a government-wide performance plan presenting a cohesive picture of the annual performance goals for the fiscal year. Agencies were required to prepare annual performance plans covering each program activity in the agency's budget. Agency performance plans were to establish annual performance goals (standards or targets) defining the level of performance to be achieved by each of the agency's program activities; briefly describe the processes and resources required to meet the performance goals; establish performance indicators to be used in measuring or assessing the relevant outputs, service levels, and outcomes of each program activity; provide a basis for comparing actual program results with the established performance goals; and describe the means to be used to verify and validate measured values. If an agency, in consultation with the Office of Management and Budget, determined that it was not feasible to express the performance goals for a particular program activity in an objective, quantifiable, and measurable form, the Office of Management and Budget was allowed to authorize an alternative form; for example, separate descriptive statements of: (1) a minimally effective program, and (2) a successful program. In preparing their performance plans, agencies were allowed to aggregate, disaggregate, or consolidate program activities, but any aggregation or consolidation was not to omit or minimize the significance of any program activity constituting a major function or operation of the agency. Performance plans may include proposals for waivers of administrative procedural requirements and controls in return for accountability for achieving specific performance goals. Performance plans may be revised to reflect congressional budget action.

Implementation of the Results Act began with a series of pilot projects in performance planning and performance reporting for fiscal years 1994, 1995, and 1996 (see National Academy of Public Administration, 1994; U.S. Office of Management and Budget, 1997a; U.S. General Accounting Office, 1997b; National Academy of Public Administration, 1998). Government-wide implementation of the statute began with agency submission of strategic plans to Congress and the Office of Management and Budget in September 1997 (see U.S. General Accounting Office, 1998a).

Agencies submitted fiscal year 1999 performance plans to the Office of Management and Budget in September 1997 and submitted final fiscal year 1999 performance plans to Congress after the submission of the President's fiscal year 1999 budget (see U.S. General Accounting Office, 1998d; U.S. General Accounting Office, 1998e). In consultation with the Office of Management and Budget, a small number of agencies chose to define some of their performance goals in an alternative, non-numerical form (see, for example, National Science Foundation, 1998, pp. 7–8; National Science Foundation, 1999, pp. 9–10). The President submitted a government wide performance plan in February 1998 as part of the fiscal year 1999 budget; the Office of Management and Budget issued a revised government-wide plan shortly thereafter (U.S. Office of Management and Budget, 1998a; U.S. Office of Management and Budget, 1998b; U.S. General Accounting Office, 1998e).

Measuring, Evaluating, and Reporting on Program Quality

Once there is a reasonable level of agreement on the meaning of program quality, policymakers, managers, staff, and other stakeholders can obtain information on key aspects of program quality through performance measurement or program evaluation. *Performance measurement* is the periodic measurement of specific program inputs, activities, outputs, intermediate outcomes, or end outcomes. Performance measurement typically focuses on collection and analysis of numerical data on agency or program performance. Performance measurement may also include qualitative assessment of agency or program performance, such as expert judgments or narrative assessments of the extent of progress toward agency or program goals.

Performance may be measured annually to improve accountability and support policy decision-making – or measured more frequently to improve management, service delivery, and program effectiveness. Performance may be measured through a variety of means including use of agency and program records, use of the records of other agencies, interviews, focus groups, surveys, and assessments by experts or trained observers. A recent General Accounting Office report describes the analytic strategies that six agencies have used to measure program outcomes, even when external factors limit agency influence over outcomes or agencies face other measurement challenges (U.S. General Accounting Office, 1998f; see also U.S. General Accounting Office, 1997a).

After studying leading public sector organizations that were working to become more results-oriented, the General Accounting Office identified characteristics common to successful hierarchies of performance indicators, noting that agencies "must balance their ideal performance measurement systems against real-world considerations such as the cost and effort involved in gathering and analyzing data" (U.S. General Accounting Office, 1996, p. 24; also see U.S. General Accounting Office, 1999). Performance indicators at each organizational level should demonstrate results, telling each organizational level the extent to which it is achieving its goals; respond to multiple priorities; and link to offices that have the responsibility for making programs work. "The number of measures for each goal at a given organizational level should be limited to the vital few" (U.S. General Accounting Office, 1996, p. 25). As agencies implement performance measurement systems, they should balance the costs of data collection against the need to ensure that the data are sufficiently timely, complete, accurate, and consistent to document performance and support decision making; in particular, that the data are sufficiently free of bias and other significant errors that would affect conclusions about the extent to which program goals have been achieved (see U.S. General Accounting Office, 1996, pp. 24–28).

Performance measurement systems should meet reasonable tests of validity, reliability, and timeliness (see Hatry, 1997; Hatry & Kopczynski, 1998, pp. 67 and 85). Though managers typically have more contextual information, both managers and policymakers need the assurance that agencies and programs have in place reasonable quality control processes that periodically review data collection procedures and test the validity and reliability of at least a sample of the data. Performance measurement systems should, therefore, be periodically reviewed and updated (Auditor General of Canada, 1997, p. 11–12 (sic); United Way of America, 1996, pp. 127–128; U.S. General Accounting Office, 1998c, p. 42). The U.S. Department of Education has developed a broad strategy for strengthening the quality of its performance data. The Department's plan includes development of department-wide standards for performance measurement; employee training in the application of the standards; standardizing definitions of key variables; coordinating data collection across different information systems; having managers, as part of their performance agreements, attest to the reliability and validity of their program performance data or submit plans for data improvement; and using audits and evaluation studies to strengthen the quality of performance data (U.S. Department of Education, 1998, pp. 87–90).

Program evaluations are systematic studies that assess how well a program is working. Evaluations are likely to be more costly and less frequent than performance measurement. Important types of program evaluation are process (or implementation) evaluation, outcome evaluation, impact evaluation, and retrospective cost-benefit and cost-effectiveness analysis. Program evaluation may examine aspects of program operations, measure difficult-to-measure outcomes, explore external factors that may impede or contribute to program success, document unintended outcomes of program activities, or measure the program's net impact by comparing program outcomes with estimates of the outcomes that would have occurred in the absence of the program (U.S. General Accounting Office, 1998b).

Program evaluation can provide a more complete, more accurate, and more credible picture of program performance than the rough sketch that is obtainable through quarterly or annual performance measurement. Evaluations can be used to identify factors that inhibit effective performance or to document more effective program approaches. Evaluations typically present options or recommendations intended to improve program performance.

No later than March 31, 2000, and March 31 of each year thereafter, the Results Act requires each agency to submit a report on program performance for the previous fiscal year. Each performance report is to set forth the performance indicators established in the agency performance plan, along with the actual performance achieved compared with the performance goals for that year – and, beginning in 2001, actual results for preceding fiscal years. Where a performance goal has not been met, the report is to explain why the goal was not met and describe plans and schedules for achieving the performance goal – or explain why the performance goal is impractical or infeasible and indicate what action is recommended. The performance report is also to describe the use and assess the effectiveness in achieving performance goals of any waiver under the statute and include the summary findings of those program evaluations completed during the fiscal year covered by the report.

Using Information on Program Quality

Performance information is intended to be useful to policymakers, managers, and other key stakeholders affected by or interested in the program. Whether in implementing the Results Act or in other performance-based management efforts, the key issue is the *use* of planning, performance measurement, program evaluation, and performance reporting: use in performance-based management systems aimed at defining and achieving agency and program goals; use in providing accountability to key stakeholders and the public; and use to support resource allocation and other policy decision making. The examples below discuss such uses of performance information.

The primary use of information on program quality should be in systems for managing agencies and programs to achieve effective performance in terms of program goals. Quality management practices include delegating authority and flexibility in return for accountability for results, creating incentives for improved program quality, reallocating resources and redirecting program activities to improve program quality, and developing partnerships designed to improve program quality (see Table 1). Many of these practices have been identified by the General Accounting Offices as common to successful efforts to become more results-oriented (see U.S. General Accounting Office, 1996, pp. 18–20, 38–46, and 51).

Table 1.

Some Quality Management Practices

1. *Delegate greater authority and flexibility in return for accountability for results*; for example, by simplifying the rules for budgeting, human resource management, or financial management.
2. *Create intangible incentives for improved program quality*; for example, through performance agreements setting challenging but realistic performance goals for agencies or programs, through quarterly or more frequent performance measurement and reporting, through meetings focused on program quality, through publicity about relative performance or changes in performance, or by delegating greater authority and flexibility in human resources manage ment or financial management.
3. *Create financial incentives for improved program quality*; for example, by introducing competition among service providers, reallocating resources to higher-performing or most-improved service providers, or linking employees' performance appraisal and pay with organizational performance.
4. *Provide training and technical assistance* to build expertise in strategic planning, performance measurement, program evaluation, and use of performance information.
5. *Redesign central management systems* (budgeting, human resource management, information management, procurement, grants management, financial management) to focus on program quality.
6. *Reallocate resources to improve program quality.*
7. *Redirect program activities to improve program quality.*
8. *Develop partnerships designed to improve program quality*; for example, partnerships among public agencies, partnerships among not-for-profit organizations, or partnerships between public-sector agencies and not-for-profit organizations or private firms.

Managers may use performance information to document the extent to which program goals have been met – and use that information to communicate the value of program activities to key stakeholders and the public. In this way, managers may use performance information to help provide accountability to key stakeholders and the public, and to support resource allocation and other policy decision making within the agency or at higher levels. In the opinions of some observers, the Results Act is beginning to change the dialogue with Congress and in the Office of Management and Budget: the way people talk about policy choice (see U.S. Office of Management and Budget, 1997b, pp. 6–7).

EXAMPLES

In 1995, the American Society for Public Administration (ASPA) initiated an effort to document experience with the use of performance measurement at federal, state, and local levels. ASPA's federal agency case studies explored the context, process, products, use and impact, and costs of strategic planning and performance measurement efforts – and identified lessons learned and next steps in strategic planning and performance measurement (see Wholey & Newcomer, 1997). Successive drafts of federal agency case studies were reviewed by the author and by Office of Management and Budget examiners prior to their publication by ASPA.

The pages below present case studies of the use of strategic planning and performance measurement to define, improve, and demonstrate quality in marine safety and environmental programs. The Coast Guard and Environmental Protection Agency (EPA) case studies show that quality management practices can be implemented effectively even in complex contexts involving regulatory programs, partnerships between the public and private sectors, and partnerships between levels of government. These and more than twenty other federal agency case studies are available from ASPA and are also available at http://www.aspanet.org and at http://www.npr.gov/initiati/mfr

Marine Safety and Environmental Protection Programs

For its pilot project under the Government Performance and Results Act, the U.S. Coast Guard developed performance plans and performance reports for its marine safety and environmental protection programs, programs "with annual spending of about $460 million and program staffing of about 3400" (U.S. Coast Guard, 1996, p. 1). Working with key stakeholders throughout the agency, the Coast Guard developed a reasonable level of agreement on outcome-related goals for reduction of deaths, injuries, and environmental damage – and developed a reasonable level of agreement on the resources, activities, and processes required to achieve the goals.

In its strategic planning process, with the participation of senior managers, the Coast Guard developed a mission statement, ". . . to protect the public, the environment, and U.S. economic interests, by preventing and mitigating marine accidents" (U.S. Coast Guard, 1996, p. 3), and established strategic goals for marine safety and environmental protection that cut across the existing organization. Using the professional judgment of senior managers, the Coast Guard then set five-year performance goals related to deaths, injuries, and environmental damage; for example:

> Reduce accidental deaths and injuries from maritime casualties by 20%. . . . Reduce the amount of oil and chemicals going into the water from maritime sources by 20% (U.S. Coast Guard, 1996, p. 5).

Senior marine safety officers from the ten Coast Guard districts validated the performance goals as reasonable and outlined strategies for achieving the goals. Given the results of stakeholder surveys and its experience in dealing with Congress, the marine industry, the states, and other agencies over the years, the Coast Guard was able to develop a reasonable level of agreement on strategic goals and on strategies to achieve the goals without the more extensive consultation now required under the Results Act (U.S. Coast Guard, 1996, pp. 2–5).

With the help of an ad hoc Program Evaluation Group from throughout the organization, the Coast Guard then developed and refined systems for assessing and reporting on performance in terms of the agreed-on goals. The Coast Guard uses data from several Coast Guard and non-Coast Guard data systems in its performance measurement systems (U.S. Coast Guard, 1996, pp. 4–7 and Attachment A).

The Coast Guard uses information from performance measurement systems and program evaluation studies to reallocate resources and redirect program activities to improve program quality – and thus improve safety, security, and the marine environment. Performance information is communicated to the field annually, quarterly, and throughout the year. As the Coast Guard reported:

> By disaggregating high-level measures like worker fatalities and passenger-vessel casualties, we have begun using trend and risk information to redirect activities and resources toward achieving our goals. . . . The main use of our performance information is to affect the outcomes themselves – to improve safety, security, and the marine environment. We do this directly by disaggregating the data to identify risk regionally and globally, in terms of . . . factors which enable targeting our efforts. . . . Our line managers in the field use the measures in our Business Plan and quarterly data extracts to help target their activities toward national goal achievement, based on local risks. . . . Managerial flexibility is inherent

(and necessary) in the process. In fact, managers can't be held accountable for achieving outcome-oriented goals without sufficient managerial flexibility to achieve those goals. This can be increased dramatically by simply reducing the organization's own internal rules and standards for activity performance (U.S. Coast Guard, 1996, pp. iii, 8, 10, and 15).

A number of agencies including the Coast Guard have created what might be termed *performance partnerships*. As the General Accounting Office reported:

[The Coast Guard] identified a significant role for the towing industry in the marine safety program and looked for opportunities to work with its stakeholders in the towing industry to reduce casualties in their field. . . . The Coast Guard and the towing industry worked to build the knowledge and skills of entry-level crew members in the industry. The Coast Guard and the towing industry jointly developed training and voluntary guidelines to reduce the causes of fatalities. This joint effort contributed to a significant decline in the reported towing industry fatality rate: from 91 per 100,000 industry employees in 1990 to 27 per 100,000 in 1995 (U.S. General Accounting Office, 1996, p. 37).

The Coast Guard has used performance information to provide accountability to key stakeholders and support policy decision making. As the agency reported:

By using outcome information in managing our programs, we met or exceeded seven of eight ambitious targets for safety and environmental protection in our first year, and five of seven in our second year, with no additional resources (U.S. Coast Guard, 1996, p. 12; also see U.S. General Accounting Office, 1996, pp. 36–37; U.S. General Accounting Office, 1997b, pp. 44–45).

Our FY95 appropriation was increased above our request by $5 million and 100 FTE [full-time equivalents] (about 1% and 3% of the totals, respectively), with Senate report language citing the importance of our program goals and the soundness of our Business Plan as the rationale. The Appropriations Committee further permitted the Coast Guard to discontinue the specific program activities associated with these resources, releasing most of these FTE for reinvestment in other activities based on safety or environmental risk, and generally endorsing increased managerial flexibility (U.S. Coast Guard, 1996, p. 12).

Chesapeake Bay Program

The Chesapeake Bay Program is an intergovernmental partnership among the U.S. EPA, the states of Maryland, Virginia, and Pennsylvania, and the District of Columbia, "built on top of the national and state environmental regulatory programs. "EPA's Chesapeake Bay Program Office manages approximately $21 million per year in federal funds and influences the allocation of substantially more federal and state resources" (U.S. Environmental Protection Agency, 1996, pp. 2–9). Working with key stakeholders, the Chesapeake Bay Program Office has developed a reasonable level of agreement on an integrated set of output and outcome goals for reductions in pollution and improvements in water quality – and developed a reasonable level of agreement on the resources, processes, and activities required to achieve the goals.

In 1987, the Chesapeake Bay Program set the core program goal of 40% nutrient reduction by the year 2000, reflecting the participants' commitment to efforts that would limit pollution from sources like wastewater treatment plants and fertilizer runoff from farmland. In the 1990s, program participants developed agreement on the program's three primary restoration objectives: reduction of nutrient enrichment effects, protection and enhancement of living resources, and reduction of adverse toxic impacts. The program uses about 30 environmental indicators to gauge the success of the restoration effort (U.S. Environmental Protection Agency, 1996, pp. 4–6 and Exhibit 2). The program measures performance in terms of outputs (administrative actions by the EPA and state regulatory agencies) and five levels of outcomes: actions by sources (e.g. installing pollution control equipment); emissions and discharge qualities of pollutants; ambient concentrations of pollutants; uptake/body burden; and health effects or ecological effects (U.S. Environmental Protection Agency, 1996, p. 6, and Attachment A, Exhibit 4).

Program staff have put special emphasis on establishing high quality systems for monitoring and evaluating the extent of progress in the Bay area. As the EPA reported:

> The Chesapeake Bay is one of the most carefully monitored bodies of water in the world. . . . Consistent and comparable data on all traditional water parameters have been taken at over 130 sites in the watershed and the open Bay since 1984. The trends data available from this monitoring program are some of the best in America. . . . During the research phase of the program, a watershed model was developed to further understanding of the Bay water quality process and the sensitivity of such processes to external nutrient loading, determined to be the main cause of the Bay's degradation. . . . Subsequent monitoring data have been used to validate this early model and to construct other simulation models used to assess the effectiveness of different pollution control strategies (U.S. Environmental Protection Agency, 1996, pp. 3-4).

The Chesapeake Bay Program has reallocated resources and redirected program activities to improve program quality. As the EPA reported:

> The availability of accepted environmental indicators has enabled the Bay Program to better target its resources. . . . Environmental indicators . . . became one of the principal methods for subcommittees to demonstrate resource needs and program success. . . . Projects from all subcommittees with nonmeasurable objectives are at a decided disadvantage against outcome-oriented projects in the contest for scarce financial resources. . . . Environmental indicators are used to develop and evaluate the effectiveness of program strategies (U.S. Environmental Protection Agency, 1996, pp. 8–9).

The program has used performance information in a variety of ways to provide accountability to the public and to support policy decision making. As the EPA reported:

> Experience has shown that public support of the program, and financial investment in the program, have been associated with the development and communication of bottom-line goals. . . . Bay Program Office staff believe that the increased support given to the program in recent years reflects the enthusiasm for supporting effective Federal-state-local partnerships to address problems. . . . [I]mprovements in the environmental indicators have facilitated goal-setting, thus better defining intended program outcomes and improving accountability to the public (U.S. Environmental Protection Agency, 1996, pp. 8–9).

CONCLUSION

As policymakers and managers work to meet increasing demands with limited resources, there is much to engage the best efforts of evaluators. Evaluators have a great deal to offer in programs directed at outcome goals, especially in hard-to-measure programs and in crosscutting programs that operate beyond the boundaries of a single agency. Program evaluation can provide a more complete, accurate, and credible picture of program performance than either day-to-day experience or the rough sketch provided by performance measurement.

Evaluators can help policymakers and managers in establishing or revising program goals, in assessing program performance, and in reallocating resources and redirecting program activities to improve program performance – and can help in communicating the value of program activities to key stakeholders and the public. In crosscutting programs, evaluators can help improve interagency coordination, helping ensure that agency goals are complementary and that program efforts are mutually reinforcing. In the Chesapeake Bay program, for example, evaluators helped policymakers in agreeing on an integrated set of goals and objectives. In the Coast Guard, evaluators helped in developing outcome measurement systems, some of which combine data from many sources and use moving averages to minimize the effects of random statistical fluctuations. Evaluators helped identify opportunities to improve program quality and program results – and helped document and communicate improved results.

In the new environment of performance-based management, evaluators can do much to improve program quality – and much to improve the quality of life on this rapidly shrinking sphere.

NOTE

1. The views and opinions expressed by the author are his own and should not be construed to be the policy or position of the General Accounting Office.

REFERENCES

Auditor General of Canada (1997). *Moving toward managing for results* (Report of the Auditor General of Canada to the House of Commons – 1997). Ottawa, Canada: Minister of Public Works and Services Canada.

Berger, P. L., & Luckmann, T. (1966). *The social construction of reality*. New York: Doubleday.

Hatry, H. P. (1997). Criteria for assessing (1) a performance monitoring process, and (2) the quality of a performance audit. Unpublished manuscript. Washington, DC: The Urban Institute.

Hatry, H. P., Fountain, J. R., Jr., Sullivan, J. M., & Kremer, L. (Eds) (1990). *Service Efforts and Accomplishments Reporting: Its Time Has Come*. Norwalk, CT: Governmental Accounting Standards Board.

Hatry, H. P., & Kopczynski, M. (1998). *Guide to program outcome monitoring for the U.S. Department of Education*. Washington, DC: U.S. Department of Education, Planning and Evaluation Service.

Joint Committee on Standards for Educational Evaluation (1994). *The program evaluation standards* (2nd ed.). Thousand Oaks, CA: Sage.

National Academy of Public Administration (1994). *Toward useful performance measurement: Lessons learned from initial pilot performance plans prepared under the Government Performance and Results Act*. Washington, DC: Author.

National Academy of Public Administration (1998). *Effective implementation of the Government Performance and Results Act*. Washington, DC: Author.

National Science Foundation (1998). *FY 1999 GPRA Performance Plan*. Washington, DC: Author.

National Science Foundation (1999). *FY 2000 GPRA Performance Plan*. Washington, DC: Author.

Organization for Economic Cooperation and Development (1996, November). *In search of results-based management practices in ten OECD countries*. Draft document prepared by the Secretariat for the Activity Meeting on Contemporary Developments in Performance Management. Paris.

Radin, B. A. (1997, September). *Performance-based management and its training implications*. Paper prepared for the International Symposium on Performance-based Management and Its Training Implications. Caserta, Italy: Graduate School of Public Administration.

S. Rep. No. 103-58, 103rd Cong., 1st Sess. Senate, Committee on Governmental Affairs. *Government Performance and Results Act of 1993* (1993).

Timpane, P. M. (1998, December 29). *Personal communication*. Washington, DC: Rand.

United Way of America (1996). *Measuring program outcomes: A practical approach*. Alexandria, VA: Author.

U.S. Coast Guard (1996). *Using outcome information to redirect programs: A case study of the Coast Guard's pilot project under the Government Performance and Results Act*. Washington, DC: American Society for Public Administration.

U.S. Department of Education (1998). *FY 1999 annual plan*. Washington, DC: Author.

U.S. Environmental Protection Agency (1996). *Use of performance information in the Chesapeake Bay program*. Washington, DC: American Society for Public Administration.

U.S. General Accounting Office (1996). *Executive guide: Effectively implementing the Government Performance and Results Act*. Washington, DC: Author.

U.S. General Accounting Office (1997a). *Managing for results: Analytic challenges in measuring performance*. Washington, DC: Author.

U.S. General Accounting Office (1997b). *The Government Performance and Results Act: 1997 governmentwide implementation will be uneven*. Washington, DC: Author.

U.S. General Accounting Office (1998a). *Managing for results: Agencies' annual performance plans can help address strategic planning challenges*. Washington, DC: Author.

U.S. General Accounting Office (1998b). *Performance measurement and evaluation*. Washington, DC: Author.

U.S. General Accounting Office (1998c). *The Results Act: An evaluator's guide to assessing agency annual performance plans*. Washington, DC: Author.

U.S. General Accounting Office (1998d). *Managing for results: An agenda to improve the usefulness of agencies' annual performance plans*. Washington, DC: Author.

U.S. General Accounting Office (1998e). *The Results Act: Assessment of the governmentwide performance plan for fiscal year 1999*. Washington, DC: Author.

U.S. General Accounting Office (1998f). *Managing for results: Measuring program results that are under limited federal control*. Washington, DC: Author.

U.S. General Accounting Office (1999). *Agency performance plans: Examples of practices that can improve usefulness to decisionmakers*. Washington, DC: Author.

U.S. Office of Management and Budget (1997a). *Report on pilot projects for performance goals*. Washington, DC: Author.

U.S. Office of Management and Budget (1997b, October). *Statement of Franklin D. Raines, Director, before the House Committee on Government Reform and Oversight*. Washington, DC: Author.

U.S. Office of Management and Budget (1998a). *Budget of the United States Government, fiscal year 1999*. Washington, DC: U.S. Government Printing Office.

U.S. Office of Management and Budget (1998b). *Government-wide performance plan, fiscal year 1999*. Washington, DC: Author.

Weiss, H. B., & Morrill, W. A. (1998, October). Useful learning for public action. Paper prepared for the Association for Public Policy Analysis and Management Annual Research Conference. New York.

Wholey, J. S. (1994). Assessing the feasibility and likely usefulness of evaluation. In: J. S. Wholey, H. P. Hatry & K. E. Newcomer (Eds), *Handbook of Practical Program Evaluation* (pp. 15–39). San Francisco: Jossey-Bass.

Wholey, J. S., & Newcomer, K. N. (1997). Clarifying goals, reporting results. In: K. N. Newcomer (Ed.), *Using performance measurement to improve public and nonprofit programs: New Directions for Evaluation, 75*, 91–98.

SECTION IV

14. EVALUATING THE AESTHETIC FACE OF LEADERSHIP

Theresa Souchet and Linda Hamman Moore

INTRODUCTION

In many ways Cheryl was a palliative to a wounded organization. The feverish brow of this school longed for a cool hand. The previous principal, Jerry, was a man given to excesses in women and drink. His small, enmeshed staff responded with warped loyalty. For the most part they closed ranks, covered for their leader, and struggled to keep their public face clean. Eventually the situation became untenable. The district office in Chicago chose a very different sort of leader in Cheryl. She appeared all-American, with her brick home in the suburbs, her three well-scrubbed children, and her devoted and successful spouse. Where Jerry had been a neglectful father to the school, Cheryl became the nurturing mother, but it was not easy taking on these guarded and cautious staff members.

Cheryl offered sustenance, bringing food to faculty meetings. They watched warily, making cutting observations that questioned her administrative exper-tise and classroom experience. She fed their egos by constantly praising their efforts, taking note of what was barely noticeable – the care with which a small bulletin board had been designed, the kind words to a wayward student, the exquisite transitions within a particular lesson. She nourished their school by working long hours, even writing grants for additional funds, funds needed in a school that had a high population of low-income students. Having inherited what many were

Visions of Quality, Volume 7, pages 219–235

ISBN: 0-7623-0771-4

calling a failing school, Cheryl worked with stalwart devotion. During the space of one year, she anticipated the availability of state funds to greatly expand innovative in-house professional development programs, and positioned her school to receive those funds. At the same time she trod carefully, taking time this first year to become sensitized to the needs of staff, waiting on the sidelines before pushing for the larger changes she thought were needed.

Cheryl's mental blueprints laid out plans for increasing trust among the staff and the children. She constructed professional development cadres, teams of teachers who would work together to support each other's scholarly and practical interests. She designed opportunities for staff members to recognize the good work of each other. Mutual appreciation and collegiality were more difficult to build among the diverse student body. Cheryl was unprepared for the level and nature of the discipline concerns at the school. Each incident acted as a door stopper, a disruption to what otherwise would have been demanding yet productive school days. As Cheryl explained, after a half year of stepping lightly, it was time to haul out stronger medicine. The previous principal may have shunned the role of "heavy", but she no longer could tiptoe around her own role as disciplinarian.

This composite portrayal of a school leader is meant to introduce the reader to some aspects of a particular principal's leadership style. The intent is to shed some light on her priorities and concerns. It is an interpretive piece, one that (on the surface) lacks "hard data". Instead, the vignette shares an overall impression. The use of metaphor throughout facilitates an understanding of the whole of her work, how she operates as a principal. While we would couple the vignette with other sources of information, we believe it could begin to tell a more complete story of this leader and her leadership style.

This representation of leadership quality differs stylistically from the personnel evaluations advocated by Scriven (1994) and Stufflebeam (1995). Both scholars have devised approaches that rely on the identification and assessment of specialized duties and competencies. Their approaches have not relied on knowledge of the literature on leadership. In this paper, in addition to reviewing the writings of personnel evaluation theorists Stufflebeam, Scriven, and others, we will discuss some of the current leadership literature that emphasizes leadership as an aesthetic process. We will argue that consideration of leadership's aesthetic face in the evaluation of quality leadership will require a shift in representation. In addition to the leadership portrayals that seem to be in vogue, evaluators should consider methods that more aptly reveal leadership's aesthetic side.

LEADERSHIP DEFINED

The study of leadership, how persons in positions of power move through their roles, is relevant to the field of personnel evaluation. Leadership as a concept has been difficult for scholars to define. A distinction has been made between being in a position of leadership, and exhibiting leadership characteristics. In our opening story about Cheryl, we mostly describe someone in the role of leader, but also attempt to comment on her working style to assist the reader in understanding her leadership qualities.

The term leadership is often coupled with a defining characteristic such as "charismatic leadership." Sergiovanni, Burlingame, Coombs and Thurston (1987) define such a leader as one who is able to provide a view of the world that creates agreements about means, ends, or both. Leadership often is distinguished from what are understood to be mundane aspects of leadership roles, such as administrative or managerial tasks. Sergiovanni et al., however, note that the distinction often serves to misrepresent the nature of leadership roles, roles that combine these different behavior modes.

Sergiovanni (1995) describes Rost's (1991) definition as a major discovery in the understanding of leadership. Leadership is explained as "an influence relationship among leaders and followers who intend real changes that reflect the purposes mutually held by both leaders and followers" (Rost & Smith, 1992, p. 195). It includes four elements: (1) the relationship is based on influence, (2) leaders and followers are the people in this relationship, (3) leaders and followers intend real change, and (4) leaders and followers develop mutual purposes. Sergiovanni emphasizes that Rost's definition captures the "interactive and multidirectional" nature of leadership. Despite presumed role differences, both leaders and followers participate in influencing each other's thinking and choices as they work toward betterment.

An important aspect of Rost's (1991) definition is that it does not simply reduce to "influence." There is a strong ethical component to leadership. Leadership is not self-serving. It is not dependent simply on a desire to fulfill bureaucratic duties. It is dependent on a shared purpose, preferably one negotiated between both leaders and followers. According to Rost and Smith (1992), leadership is based on credibility, which encompasses character, courage, competence, composure, and caring. Trusting and honest relationships are vital for leadership, relationships in which persons are willing to stand up for their beliefs, trust in the ability of others, compose themselves graciously, and have concern for the welfare of others.

Rost's (1991) definition explicitly moves away from the fulfillment of duties and tasks in defining leadership. Although Sergiovanni et al. (1987) make clear that managerial duties are aspects of the role that should not be diminished, they adhere to Rost's assertion that leadership is much more. In evaluating persons in positions of leadership, Stufflebeam (1995), Scriven (1994) and others do not appear to be interested in distinguishing between the different aspects of leadership, or even the distinction between leadership and persons in positions of leadership. Instead they focus on the tasks which comprise the domain of leadership roles. Some of the tasks may rely on the same qualities included in Rost's definition (e.g. character, courage, and caring), but it is questionable whether the methods they advocate to evaluate and to represent persons in positions of leadership adequately tap this aspect of leadership.

We describe the work of evaluation theorists such as Scriven (1994) and Stufflebeam (1995) as rational-analytic. The approaches rely heavily on evaluation of leaders based on a delineation and description of duties that comprise their roles. Other theorists, such as Heck and Marcoulides (1996) and Duke (1992), are more interested in effects and outcomes that can be linked to persons in position of leadership. What follows is a discussion of their work.

RATIONAL-ANALYTICAL APPROACHES TO EVALUATING LEADERSHIP

Duties-based Evaluation

In attempting to evaluate and represent personnel, some evaluation theorists have worked diligently to develop extensive, detailed lists of job duties and standards. Although Scriven (1994) has not developed an approach to the evaluation of leaders per se, his *duties-based approach* to personnel evaluation could be used to evaluate educational leaders. His "Duties of the Teacher" (DOTT) list is an attempt to describe comprehensively the generic duties of teaching which "distinguish it from the work of other professions" (p. 153). A number of subareas and subelements cluster under five domains: knowledge of subject matter, instructional competence, assessment competence, professionalism, and other duties to the school and community.

Anticipating criticism that he is losing meaning in reducing the complexities of teaching to a list of duties, Scriven (1994) offers the list as a way of beginning to understand those complexities. He asserts, however, that the duties-based teacher evaluation model is the only legitimate basis for evaluating teachers, unless a particular teacher's system chooses to apply stylistic or current "fad" criteria in making judgments, obviously not a practice that he endorses.

Scriven (1994) provides an example of how an evaluator might use a graphical representation of an individual teacher's performance, applying local standards of quality to the DOTT list. The representation appears as a bar graph, with the five domains listed across the horizontal axis and five letter grades (A, B, C, D, and F) on the vertical axis. Additional pages can be attached to include comments by the evaluator or by the teacher, or to analyze more specifically areas needing remediation. Assuming that these attached pages contain narratives, we suspect that they are more likely to help us understand the unique qualities of an individual than the bar graph.

Minimizing any reference to "style" of teacher, Scriven (1994) advances his DOTT list as "a *normative* list . . . of what teachers can *legitimately* be held responsible for knowing and doing" (p. 156). He is right, of course, that this reductionist approach is open to criticism. While the list seems exhaustive – and exhausting – a list of duties cannot capture the whole of a teacher's quality any more than the competency lists and matrices of the aesthetics proponents we will discuss later in this paper.

Standards-based Evaluation

Stufflebeam (1995) extends Scriven's (1994) work by advocating a *standards-based model* for the evaluation of superintendents. He is eager for evaluators to reach consensus on a common evaluation model. He describes the work of the American Association of School Administrators (AASA), which has published a set of superintendent competencies describing the knowledge and skills for superintendent preparation, certification, and professional development. He supports the work of University of Texas researchers in developing the Diagnostic Executive Competency Assessment System (DECAS), based on a hierarchy of leadership domains, tasks, and competencies. In agreement with Scriven's approach described above, Stufflebeam worked with the Texas Education Agency to develop a list of generic superintendent duties. Further, he states that there is growing agreement that evaluations of schools, teachers, districts, and administrators should be based, in some way, on student achievement measures.

Stufflebeam and Millman (1995) propose a superintendent evaluation model to address what they perceive as deficits of existing models (see Candoli, Cullen, & Stufflebeam, 1995, for a summary and categorization of existing models). They suggest that there are four general tasks in assessing superintendent merit or worth. *Delineating* tasks relate to decisions made between the superintendent and the board about the evaluation: what it will address, which audiences have access to which results, what are intended uses, how various accountabilities are weighted, and what standards will be used. *Obtaining* tasks refer to the collecting and analyzing of data. Types of data collected fall under the familiar Stufflebeam model of Context, Inputs, Process, and Products (CIPP). Data are used to address the evaluation questions developed during the delineating tasks. The reporting of information to the agreed-upon users is a *providing* task. Providing tasks can be either formative or summative. *Applying* tasks refer to the uses of the evaluation. In Stufflebeam and Millman's model, evaluation reports should be used to make decisions about the superintendent and the district, and to guide the board's communications with the media and community.

Stufflebeam and Millman (1995) have developed a matrix integrating the AASA superintendent competencies with the superintendent duties developed in Texas. They encourage all school districts to adopt a generic set of duties to be used in evaluating superintendents. There are 11 general duties proposed by Stufflebeam and Millman, with many specific duties listed to illustrate the general. They consider numerous ways to represent the superintendent quality at each stage of the CIPP evaluation, including stakeholder panels, climate surveys, principal reports, hearings, data analyses of various sorts, portfolios, and synthesizing reports. They even suggest how tasks might be divided over the four quarters of the calendar year, and how different types of evaluation (C, I, P, or P) might be emphasized differently during different quarters. In any event, all aspects of the evaluation are to be driven by communication between the evaluator and the superintendent, the generic duties of the superintendent, and adherence to the *Personnel Evaluation Standards*. While the flexibility of Stufflebeam and Millman's solution seems to allow something for everyone, we wonder about the feasibility of these potentially complicated and expensive superintendent evaluations–particularly in smaller school districts that are not resource rich.

Role-based Evaluation

Heck and Marcoulides (1996) report a method of evaluation that will not be so easily understood or implemented by a board. While they see duties- and standards-based evaluations as useful beginnings, they are more interested in the empirical effects of administrative actions. They refer to their approach as *role-based evaluation*, as they study what are considered important effects of a certain role's (e.g. a principal's, in this study) activities. They assert that public demands for accountability, cost containment, and productivity mandate the use of "psychometrically and sound observations of performance" (p. 26).

Using a two-level hierarchical linear model, they explored both within- and between-schools differences to compare how assessments of principal leadership vary with the role of the assessor (teacher vs. principal), the hierarchical nature of the data, and school features and contexts (high school vs. elementary, high- vs. low-achieving school). They controlled for socioeconomic status and student language background. A 34-item questionnaire was given to principals and teachers. From their earlier research they expected three dimensions of principal leadership to emerge from the data: leadership activity related to governance, activity to improve school climate and social relations, and activity toward monitoring instruction. These dimensions were confirmed in their factor analysis.

After analyzing sources of between- and within-school variance, Heck and Marcoulides (1996) ultimately argue against between-school analyses of leadership, explaining that they may violate standards of propriety and accuracy. It is interesting to see their endorsement of advanced psychometrics concurrent with their support of particularization. They endorse flexible, multidimensional evaluation models that include multiple data sources, allowing various critical perspectives on the leader's role to be represented. Although their approach also falls under the rational-analytic heading, they seem far from endorsing lists of generic standards.

Administrative Effectiveness

Duke (1992) attempts to clarify the struggles evaluators have faced in determining administrative effectiveness. He identifies four different concepts of administrative effectiveness: effectiveness as a function of traits, compliance, competence, and attained school outcomes. Each concept of effectiveness, he proposes, has different criteria for determining purpose, validity, measurement, and administrative image. For example, evaluations may be conducted for purposes of accountability, control, selection, advancement, or professional development.

When effectiveness is seen as a function of compliance, it has accountability as its main purpose. A duties-based evaluation, such as described by Scriven (1994), seems to fit within this category, although Scriven also clearly is interested in evaluating teacher competence. Scriven's compliance interests appear in his description of the DOTT list as what teachers "should be doing" (p. 156) and in his admonishment to teachers that they are to be accountable to themselves and to third parties. Leadership, according to Duke (1992), implies initiative, but seeing compliance as the indicator of effectiveness gives weight to management over initiative.

Standards-based evaluations, such as those endorsed by Stufflebeam (1995) prioritize effectiveness as competence. Performance standards, by definition, imply an emphasis on assessing at least minimum competencies. Although accountability purposes seem primary, purposes of prediction, selection, and professional development also can be addressed. Key performance areas (that we might consider part of the aesthetic face of leadership), such as professional judgment, or the ability to anticipate problems, are given little attention. While the use of standards seems to "make sense," it is important to consider that standards are developed, and performance levels are judged, within a political process, and not within the mythical, 100% rational process they try to serve. Because of this, evaluators must maintain some level of skepticism about standards' face validity.

Accountability is served again when effectiveness is viewed as the function of attained school outcomes. Like many evaluators, Duke (1992) is troubled by the validity of basing administrative evaluations solely on school outcome measures. He speculates that outcome data more often are used to support judgments actually based on other concerns. For this reason, some evaluators would omit any use of school outcome measures in evaluating the quality of school administrators. Certainly, school outcome measures are the result of incredibly complex, higher order interactions. It could be argued that the quality of the educational leader is one factor in the interaction, but it does not follow that school outcome measures reflect the quality of the leader.

Finally, Duke (1992) states that the purpose of assessing effectiveness as a function of traits is selection, or deciding who is administrator material. A leader's traits, however, may exert "a certain generalized or 'aesthetic' effect on subordinates and clients" (p. 105). Because traits, by definition, are stable and enduring personality constructs, however, he questions the need for ongoing evaluation of them. Duke proposes a fifth concept of effectiveness, administrative outcomes, which are the desired consequences of a completed administrative task. He sees these administrative outcomes as mediating variables between task compliance and school outcomes.

Duke (1992) emphasizes the need to consider specific purposes of personnel evaluation. It is likely that evaluations will be required to fulfill distinct needs, needs not fulfilled by the standards-based or duties-based models presented. When one is assessing whether a person is "leadership material," evaluations may need to consider more carefully what scholars in the field are saying about the character of leadership.

All of these evaluators understand that the analytical approaches they favor are, nevertheless, "embedded in social dialogue" (Palus & Horth, 1996, p. 57). Even while attempting to generate a list of typical duties, Scriven (1994) discusses applying local standards of quality. While Stufflebeam (1995) hopes for a consensus on standards, he also describes the necessary communications between the evaluator and the individual being evaluated. Also, Heck and Marcoulides (1996) plainly argue for interpreting statistical data within the context of an individual school when the data are being used to evaluate leadership quality. None of the authors we have discussed under Rational-Analytic approaches, however, shows an appreciation for the connection between aesthetics and leadership.

THE AESTHETIC FACE OF LEADERSHIP

The study of leadership has been combined with the study of aesthetics in a 1996 volume of *The Journal of Aesthetic Education*. Several authors, including Smith (1996), Howard (1996), and Moore (1996), discuss aesthetics and its relationship to leadership practice and process. The discussions are complex, drawing from the ideas of ancient and modern philosophers. Taken together, the papers argue for increased attention to the contribution of aesthetics, the philosophy of art.

In bringing together these two bodies of literature, aesthetics and leadership, the journal's authors adopt several approaches. Some papers approach this line of inquiry by discussing the contribution of aesthetic education to leadership training. Others talk about developing aesthetic competencies. Still others argue for a more integrated approach to the study of the aesthetic side to leadership, one that views leadership as a fundamentally aesthetic process. In the following section, we present several scholars grappling with the link between leadership and aesthetics.

Aesthetics as Competencies

Noticing, "slowing down and taking in more of an image, a scene, or a situation" (Palus & Horth, 1996, p. 60) is an aesthetic competency, as is *subtle representation*, the ability to portray with subtlety, detail and accuracy that which has been apprehended. Other aesthetic competencies include *fluid perspective* (perception which encompasses multiple perspectives, and changes with time); *personalizing work* (allowing artistic interests, or processes that support these interests, to spill into one's job); and *skeptical inquiry* (asking questions that add insight by fostering doubts and alternative viewpoints). The significance of these competencies, according to Palus and Horth, is that they allow for a language of leadership that is "based in the senses, perception, imagination, desire and meaning making" (p. 64). These authors argue implicitly against a theory of leadership based solely on rational/analytic adeptness, a focus of a number of evaluation theorists. Aesthetic competencies are meant to provide a "new" language, one that embraces leadership as art making. The authors suggest that these competencies can assist leaders in re-defining their roles, but caution against the use of these competencies as a training checklist. Rather, they see their list of creative problem-solving techniques as necessary but insufficient for "sustaining creative leadership" (p. 65).

The competencies presented by Palus and Horth (1996) include a wide range of abilities. The competencies are a good starting place from which to consider the meanings of aesthetic leadership. The authors are sensitive to the use of the word "competency", noting that it is itself perhaps, "too mechanical, too analytical, too small" (p. 64). We suggest that the form of their analysis, a list of competencies, is useful in some respects, but is itself overly mechanical, analytical, and limiting. Palus and Horth find it important to steer the conversation on leadership in a different direction, one that considers the complex ways in which leaders can be encouraged to work creatively. In doing so, they advocate consideration of aesthetic competencies as a way for leaders to pay less attention to formal boundaries of work (e.g. standard operating procedures and mission statements) and more attention to what is difficult to apprehend. Disciplined aesthetic sensibilities, they argue, are part of learning to resolve unknown spaces into larger "patterns, movements, and interconnections" (p. 64). Their ideas succeed in redirecting the conversation. Attempting to represent aesthetic leadership with a list of competencies, however, seems little more than another analytical trait-based approach, one that we criticize Scriven (1994) for using in evaluating personnel.

A more extreme example, it seems, is found with Merritt and DeGraff (1996), who attempt to measure leaders' levels of aesthetic awareness across eight domains. They offer a 5 × 8 matrix combining the five aesthetic developmental stages of Parsons (identified by assumptions individuals tend to have when confronted with works of art) with the eight dimensions – "underlying values, reasoning, experience, or assumptions out of which the behaviors of extraordinary leaders grow" (p. 77) – from Thompson's (as cited in Merritt & DeGraff, 1996) Visionary Leadership Inventory. They write:

> We need to locate the leaders on the awareness table, design suitable measures, and relate assessed outcomes back to the domains of acquired competencies. This model of described competencies is a preliminary sketch of what leaders in any given domain might recognize as an indication of their levels of aesthetic development (p. 81).

It seems peculiar to use a competency-based model such as this while offering rhetoric about the integration of the senses, imagination, and intellect. Although the list of competencies described by Scriven (1994), Palus and Horth (1996), and Merritt and DeGraff (1996) differ in form and content, we believe that this kind of approach to the analysis of leadership fails to attend adequately to the complexity of quality leadership.

How Aesthetic Theory Informs Leadership Study

Moore (1996) approaches the contribution of aesthetic theory to leadership and other "worlds of action" (p. 7) differently. Concentrating on how aesthetic theory informs leadership study, he outlines several theories and explains how they inform leadership practice. The *institutionalism* theory of art is one such line of inquiry. The institutionalism theory underscores meaning making as a social process. It also emphasizes the idea that judgment in the art world comes out of a network of people and histories. The theory is important, says Moore, because its implications necessitate an understanding of where an organization was, is, and can be.

A second theory, *aspectivism*, is the idea that "a thing's value is related to the multiplicity of ways in which it can be perceived" (Moore, 1996, p. 13). The value of a piece of artwork is somewhat dependent on the multiple interpretations that can be brought to bear on the work. According to Moore, this idea is important in that it makes us aware of the dexterity of our own consciousness. Cognitive dexterity, he states, leads to the ability to imagine, an important capacity for leaders: "One of the worst enemies of successful leadership in any field is tunnel vision, the persistent tendency to look at things in just one way" (p. 13).

Another "tendency of thought" Moore (1996) discusses is *aesthetic ecumenism*. It is defined as "a mode of thought [which] seeks to link people of our time and place with people from other times and places through a shared quality of some of their aesthetic experiences and perceptions" (p. 16). Moore suggests that fundamental commonalities of human experience are revealed in great works of art. As the consumer experiences great art, "imaginative sympathy" allows for connections between the creators and others who share the experience. This coming together in shared experiences is viewed as an important way to increase understanding and bridge differences among members of an increasingly shrinking and interconnected world.

Listed here are just a few of the aesthetic theories and thought patterns discussed by Moore (1996). In delineating these theories Moore provides a useful set of considerations for examining the possible influence of aesthetic theory on leadership. These thought processes differ from Palus and Horth's (1996) competencies in that they lean toward descriptive rather than normative characteristics. Instead of solely defining what leadership should be, Moore describes the ways in which aesthetic theory can inform leadership practice. Moore emphasizes the value of such patterns of thought. He compares his message to that of Cicero. Rather than becoming philosophers (in Moore's case, "aestheticians") Cicero "taught that philosophy and reflective acquaintance with the arts were the best available means of cultivating the required practical skills in the mind of the leader" (p. 18).

Leadership as an Aesthetic Process

Howard (1996) takes ideas about aesthetic leadership one step further. Whereas Palus and Horth (1996) and Moore (1996) mine theories of art for ways in which leadership is influenced by and reflective of aesthetic processes, Howard argues that leadership is a fully aesthetic experience. Howard coined the term "the aesthetic face of leadership" to refer to "that elusive amalgam of perceptive sensitivity (sensibility), judgment, persuasiveness, imagination, and timing – much like that of an actor or performing artist – exemplified in the actions, decisions, rhetoric, and public presence of acknowledged great leaders" (p. 25).

Howard's arguments grew out of Dewey's (1934) distinction between *episodes* or *situations* and *an experience*. An experience, Dewey writes, "has a unity that gives it its name, *that* meal, that storm, that rupture of friendship. The existence of this unity is constituted by a single quality that pervades the entire experience" (p. 37). Howard labels that quality "emotionalized thinking", and when such a quality pervades an event, says Dewey, the experience bears an aesthetic stamp. According to Howard, emotionalized thinking constitutes the work of leaders. Howard agrees with Wills (as cited in Howard, 1996) that a leader is an "interpreter of followers' feelings and thoughts" (p. 26). The process is complex, Howard notes, as it "involves the full engagement of one's sensibilities in successive reconstructions under ever-changing, often highly stressful conditions" (p. 27). It is the engagement of sensibilities under unstable circumstances that requires emotionalized thinking, stamping the process of leadership as an aesthetic one.

Although his approach is different, Smith (1996), like Howard (1996), views leadership as an aesthetic process. Smith's definition of leadership relies less on influence over others and more on meaning making. He follows the views of Drath and Palus (as cited in Smith, 1996), who construe leadership as "the efforts of the members of a community of practice to achieve a common understanding of a difficult task" (p. 39). In working toward common understanding, Smith fully compares the leader to the artist, explaining that the "product" is "the quality of the performance by the members of a corporate unit in their endeavor to achieve common understanding . . . [The] performance implies both the process of striving and its result" (p. 43). It is this process of striving that can make an aesthetic presentation. Communication across persons with differing dispositions, values, ideas and backgrounds, also affected by their positions within an organization, is complex. Smith calls such communication *interdisciplinary inquiry*. He argues that an education in aesthetics enhances the ability to communicate across differences, partly because of the training in the use of metaphor. The use of metaphor is significant, in that similarities are revealed between different phenomena. More generally, Smith says the sharpened perceptiveness and discrimination cultivated by a study of the arts will, among other things, "foster sensitivity to the emotional properties of situations" (p. 47). Such sensitivity is important in communicating across differences.

Smith (1996) is suggesting that leaders, like artists, must tap their aesthetic side when engaged in their craft. What is required of leaders is successful application of their aesthetic sensibilities. In this sense he is agreeing with the notion of aesthetic competencies, but failing to separate them from the act of leadership, presumably quality leadership. Smith also points out the importance of engagement in metaphor, a tool we believe is important in the representation of leaders in leadership roles. Langer (1957) earlier described metaphor as "the natural instrument of abstract thinking" (p. 104). She points out that we know and understand more than our discourse. We recognize abstractions in our subjective experience before we have words to label those abstractions, so we rely on metaphor. One might say that the aesthetic face of leadership itself is such an abstraction, and that it is a challenge to represent such a thing with words. Metaphor helps as we attempt to portray that which resists being represented by discourse. Cheryl, the principal in the opening vignette, may not be a soothing remedy, a nurturing mother to the school, nor an ambitious architect. Metaphorical language, however, conveys the quality of aspects of her leadership more understandably than a checklist of competencies.

The literature on aestheticism and leadership labels as aesthetic those leadership capacities considered by many to be intangible. The capacity to see different points of view, to take notice of that which is difficult to perceive, and to appreciate the contribution of metaphor to communication are some of the qualities mentioned. One would be misdirected, however, in viewing the task of these scholars as simply one of categorizing, or of distinguishing aesthetic competency from rational-analytic capacity. Several authors in *The Journal of Aesthetic Education* (1996) are devoted to improving the quality of leadership. Their interest is as much in identifying what is aesthetic about leadership as it is in arguing that good leaders access rational-analytic skills alongside aesthetic ones. The volume can be viewed as an attempt to get a handle on the aspect of leadership that is difficult to quantify and measure, an aspect that has been alluded to but (we believe) not adequately addressed by personnel evaluation theorists.

REPRESENTING LEADERSHIP'S AESTHETIC SIDE

We have shown that some scholars of leadership and evaluation, including those writing of the connection between leadership and aesthetics, turn to rational-analytic approaches to study leadership and to evaluate those in leadership roles. Many involved in this research also recognize an intangible nature to the study of leadership, and note the limitations in using one method of analysis. Others advocate a turn to in-depth study of leadership, one that includes the study of leaders in context. Howard (1996), for instance, argues that the better studies of leadership phenomena take full advantage of naturalistic method, displaying "a marked preference for anecdote and portraiture to fill out and support generalizations [about leadership]" (p. 29). He points out that the strength of what he calls "humanistic research" (methods used by historians, biographers, and journalists) is that it can dwell on personal and circumstantial details. These details often are underplayed by analytical approaches. Howard asserts that studies which examine the particulars in the lives of leaders are well suited to examining the aesthetic face of events in leadership, and that "ignoring [these events], however rigorous one's theoretical formulations and quantitative data, is to miss the heart of the matter" (p. 32).

What we want to add to Howard's (1996) critique is an emphasis on the importance of representation in leadership studies, whether they entail evaluation of leaders, or the more general study of leadership. We began this paper with a short narrative, attempting to employ techniques that provide an image of a leader. Many would not feel satisfied with our portrayal, arguing that even within the qualitative genre it is "too interpretive," and in this sense "extreme." Our intent is to draw the reader to a style that moves beyond more accepted narrative tellings (i.e. those purporting to rely more heavily on objective description). The narrative, we hope, clearly brings to light our own understanding of this leader. In taking full advantage of metaphorical language, we make an effort to capture what is abstract about this leader's style. To give the reader a more detailed sense of this principal's leadership, we might add a portrayal of an interaction between the principal and colleagues or teachers, or describe a significant event this principal was engaged in.

Narrative as a form of representation is advocated by some scholars in sociology and education. Richardson (1997), for example, writes that narratives are important in helping people make connections between data: "It is the primary way through which humans organize their experiences into temporally meaningful episodes" (p. 27). Narrative is a familiar way to understand and learn:

> If we wish to understand the deepest and most universal of human experiences, if we wish our work to be faithful to the lived experiences of people, if we wish for a union between poetics and science . . . then we should value the narrative (p. 35).

There are serious issues to address, of course. Perhaps most important for evaluators is how to judge the legitimacy of narrative representations of leaders and other personnel. Scholars advocating the use of narratives have sought answers in different arenas. Tierney (1993) suggests a turn to literary criticism; Denzin (1997) turns to poststructural forms of legitimation, for instance, "politics." These solutions are not likely to be satisfactory to the evaluation community. The realities of evaluation, such as its role in helping with policy decisions, are such that there is a need to verify representations. Despite an affinity we all share for storytelling, its use in the realm of personnel evaluation will be questioned, if not resisted entirely.

It may be useful to remind readers of the early resistance to (some might say outright rejection of) qualitative methods in evaluation. Scholars, such as Guba and Lincoln (1989), and Stake (1995), fought uphill battles in convincing others of the worthwhile contributions of naturalistic design. The refinement of qualitative technique, the development of "organizers" to discipline inquiry (e.g. issues), the use of confirmatory strategies and peer critique, the reliance on multiple interpretations, and the emphasis on the insertion of the researcher's identity into reports, have resulted in greater acceptance. The techniques are understood as challenging and rigorous, and demanding of one's abilities to synthesize, integrate, and enlighten.

Representations of evaluands vary in the degree to which the strengths of qualitative technique are reflected. Arguments for the use of qualitative methods frequently hinge on beliefs that they are well suited to allowing a different way of seeing "the object of study." To exploit the strengths of naturalistic method, it seems logical to explore different methods of representation. This is, we feel, an important role of the evaluator. If the task is to assess quality, then the evaluator should be versatile in using methods and, equally important, in representing "findings."

Quality is not a static characteristic of an evaluand, in this case a leader. As evaluation theorists recognize, it is informed by values, contexts, roles, and perceptions of various players (stakeholders). Quality is open to the interpretation of evaluators making the effort to understand the process of leadership. In assessing leaders, some evaluators rely on the delineation of broadly defined duties. There is agreement that the assessment of leaders is imbedded in contexts (including the roles of those involved) and dependent on purpose. When assessing leadership, however, there is a need to move beyond role expectations. What the research on leadership emphasizes is that all persons in positions of leadership are not leaders. Authors such as Moore (1996), Smith (1996), and Howard (1996) point to what might very well be needed to distinguish leaders from persons merely in a position of leadership. Perhaps quality leaders have developed their aesthetic side, including an understanding of processes of thought that match those identified by Moore. Or perhaps they have come to hone their successes as they move through the aesthetic moments described by Howard. This is not to say that leaders are conscious of this interpretation of their leadership. We do think, however, that evaluators should be aware of aspects of leadership difficult to ascertain with conventional forms of data collection (e.g. student test scores or performance surveys). Evaluators cognizant of the "aesthetic face of leadership" will have another lens from which to study, understand, and represent the leadership quality.

REFERENCES

Candoli, C., Cullen, K., & Stufflebeam, D. L. (1995). Analysis of alternative evaluation models. *Journal of Personnel Evaluation in Education*, *9*, 369–381.

Denzin, N. K. (1997). *Interpretive ethnography: Ethnographic practices for the 21st century*. Thousand Oaks, CA: Sage.

Dewey, J. (1934). *Art as experience*. New York: Milton, Balch, and Company.

Duke, D. L. (1992). Concepts of administrative effectiveness and the evaluation of school administrators. *Journal of Personnel Evaluation in Education*, *6*, 103–121.

Guba, E. G., & Lincoln, Y. S. (1989). *Fourth generation of evaluation*. Newbury Park, CA: Sage.

Heck, R. H., & Marcoulides, G. A. (1996). The assessment of principal performance: A multilevel evaluation approach. *Journal of Personnel Evaluation in Education*, *10*, 11–28.

Howard, V. A. (1996). The aesthetic face of leadership. *Journal of Aesthetic Education*, *30*, 21–37.

Langer, S. K. (1957). Problems of art: *Ten philosophical lectures*. New York: Charles Scribner's Sons.

Merritt, S., & DeGraff, J. (1996). The revisionary visionary: Leadership and the aesthetics of adaptability. *Journal of Aesthetic Education*, *30*, 69–85.

Moore, R. (1996). The nightmare science and its daytime uses. *Journal of Aesthetic Education*, *30*, 5–20.

Palus, C. J., & Horth, D. M. (1996). Leading creatively: The art of making sense. *Journal of Aesthetic Education*, *30*, 53–68.

Richardson, L. (1997). *Fields of play: Constructing an academic life*. New Brunswick, NJ: Rutgers University.

Rost, J. C. (1991). *Leadership for the twenty-first century*. New York: Praeger.

Rost, J. C., & Smith, A. (1992). Leadership: A postindustrial approach. *European Management Journal*, *10*(2), 193–201.

Scriven, M. (1994). Duties of the teacher. *Journal of Personnel Evaluation in Education*, *8*, 151–184.

Sergiovanni, T. J., Burlingame, M. B., Coombs, F. S., & Thurston, P. W. (1987). *Educational Governance and Administration*. (2nd ed.). Englewood Cliffs, NJ: Prentice–Hall.

Sergiovanni, T. J. (1995). *The Principalship: A reflective practice perspective*. Boston: Allyn and Bacon.

Smith, R. A. (1996). Leadership as aesthetic process. *Journal of Aesthetic Education*, *30* (4), 39–52.

Stake, R. E. (1995). *The art of case study research*. Thousand Oaks, CA: Sage Publications, Inc.

Stufflebeam, D. (1995). A conceptual framework for study of superintendent evaluation. *Journal of Personnel Evaluation in Education*, *9*, 317–333.

Stufflebeam, D., & Millman, J. (1995). A proposed model for superintendent evaluation. *Journal of Personnel Evaluation in Education*, *9*, 383–410.

The Joint Committee on Standards for Educational Evaluation. (1988). *The personnel evaluation standards*. Thousand Oaks, CA: Sage.

The Journal of Aesthetic Education, Vol. 30, Number 4, Winter 1996 (Special issue: the Aesthetic Face of Leadership), University of Illinois Press.

Tierney, W. G. (1993). The cedar closet. *Qualitative Studies in Education*, *6*, 303–314.

15. THE METAPHOR OF THE PORTFOLIO IN EVALUATING TEACHER EDUCATION PROGRAMS

Judith Walker de Felix and Alexis P. Benson

INTRODUCTION

The portfolio is becoming widely used for evaluating students and their educational programs. Nowhere is this more evident than in teacher education, as increasing numbers of state departments of education specify standards for teacher certification and then evaluate those programs by examining evidence that the standards have been met. That evidence typically includes portfolios that document pre-service teachers' achievements. State agencies then judge teacher preparation programs as acceptable or not based on multiple measures of participants' skills, including those documented in their portfolios.

The objective of this chapter is to determine the role that portfolios might play in going beyond such evaluation to achieve a more data-rich description of the quality of programs. The paper first traces how educators implemented portfolios in two very different teacher education programs. Next, the paper will review the research literature that describes the role of portfolios as documentation of the performance of pre-service teachers. It will then synthesize the findings in light of metaphors that characterize how teacher educators employ portfolios. Finally, the conclusion will suggest how portfolios might

Visions of Quality, Volume 7, pages 237–251

ISBN: 0-7623-0771-4

allow evaluators to view the quality of teacher preparation programs from the stakeholders' perspectives.

Because students are one of the major stakeholder groups, evaluating the quality of a teacher education program requires taking into account their perspectives. Academic portfolios are powerful evaluation tools because they depict students' experiences in the teacher education program. They are designed to be a means for students to exhibit learning over time. They allow readers to make a broad-based, authentic evaluation from a variety of data sources and data-collection methods. The portfolio contains longitudinal evidence of the student's efforts to improve. As such, a portfolio is intended to reflect development (Touzel, 1993).

All assessments reflect assumptions about knowledge and competence. Traditional tests are rooted in the belief that knowledge and skill can be characterized in terms of separate bits of mental associations. On the other hand, alternative assessments, such as portfolios, are based on theories of situated cognition. In situated cognition the state of the learner's preparedness interacts with the tools, people, and context to produce performance (Resnick, 1996).

As teacher preparation programs move to authentic experiences in schools, pre-service teacher assessment also needs to become more authentic. The portfolio has become the preferred means for providing authentic assessment.

PORTFOLIO USE IN PRACTICE

This section describes the portfolio development and implementation process in two teacher education programs. The two sites illustrate not only the complexity of teacher preparation programs but also how portfolios can shed light on the quality of two very different programs.

Site A

Site A is in a public Carnegie Research II university with several graduate programs, including doctoral programs in various areas of education. The teacher education program admitted about 450 new students per year in a state with very prescriptive requirements for teacher certification. Site A has been accredited by the National Council for the Accreditation of Teacher Education (NCATE) for most of its history, and many of its faculty have participated on NCATE teams.

Site A had been developing a teacher education program based entirely in schools for seven years. They had several documents, including electronic information, describing their program and assessment expectations. Their program descriptions, for example, cited scholars in situated cognition and neo-Vygotskian pedagogy (Brown, Collins & Duguid, 1989; Cognition and Technology Group at Vanderbilt, 1990; Tharp & Gallimore, 1988) and claimed that the program relied on authentic tasks that minimize any inert knowledge, i.e. declarative knowledge that is not anchored to a personally and socially relevant activity. Their stated goal was for their graduates to have a sense of empowerment developed through meaningful professional experiences.

Descriptions of the use of portfolios were also very complete at Site A. Faculty claimed that portfolios require students to assemble evidence of milestones they have reached. This allowed students from diverse backgrounds to use their own "voices" and products to make the case for evidence of their learning. Faculty guided students to use a "value-added" principle in assembling documentation. That is, students were not to add evidence that did not increase the value of their portfolio. This prevented interns from thinking that a bigger portfolio is better.

Site A also relied on research to guide their utilization of portfolio information. For example, they subscribed to Nweke's (1991) findings that portfolios supplement rather than substitute for traditional testing procedures. Traditional examinations in campus-based core courses and the state-required certification examinations continued, but courses with significant experiences in schools were assessed with a portfolio.

At Site A portfolios were introduced during the application procedure. Applicants had to assemble evidence that they were ready for the professional development phase of their preparation program before being allowed to enter the first education course. Because students were going into the schools earlier, the faculty wanted to assure that the pre-service teachers had a professional attitude. They asked applicants to supply evidence that they had experience working with young people, interacting with others from diverse cultural and linguistic backgrounds, and reflecting on teaching as a career. Applicants also had to provide letters of reference with phone numbers for admissions counselors to verify the information provided. One letter was required from a faculty member who had agreed to mentor the applicant. In addition to the expanded requirements, the students also had to fulfill criteria similar to those in the previous traditional program: grades, lower division courses completed, and state-required courses.

Site A students' next portfolio was required after students had observed teachers, taught lessons to small groups, and served as teachers' aides in a professional development school (PDS). University faculty, P-12 teachers, and the university students discussed ways in which the pre-service teachers could document that they were ready to student teach. Students often included products that they developed as they solved the problems posed by university faculty in their PDS courses. Students then prepared portfolio presentations similar to graduate students' theses defenses. Each student assembled a committee of university and school faculty to hear the evidence that the candidates had met the objectives of the second phase of the program.

The final portfolio was presented when the student teachers believed they had completed student teaching. At this time, candidates had to demonstrate that they were "safe to start" teaching. The same committee again met, listened to the presentation, asked questions, and made recommendations based on the final portfolio.

During the last semester, faculty also invited principals and personnel directors to present their views on evidence that they would like to see at job interviews. Candidates began to discuss among themselves and with school and university faculty the items in their portfolios that they could extract to include in a professional interview portfolio.

The practices at Site A reflected its long history of a team approach to teacher preparation. The Associate Dean for Teacher Education negotiated roles with faculty and their department chairs and assigned teaching assistants. Faculty members designed group review processes for providing feedback on the portfolios to minimize the time that they spent on each portfolio. The team also secured external funding to reward faculty with technology and research assistants. The administration facilitated research plans so that faculty could meet their own career goals. For some faculty participants, the portfolio provided data for that research, including studies of the university students' concept maps (Shavelson, Lang & Lewin, 1994).

Site B

Site B is at an independent liberal arts university that admits about 55 new teacher education students per year. Site B is classified as Carnegie Comprehensive, and there are masters and educational specialist degrees in the School of Education. The state certification requirements for Site B are in transition from a traditional input model to one requiring outcomes based on modifications of the NCATE standards. Site B has resisted NCATE historically but is considering applying for accreditation because the state is moving toward a single accreditation model.

Site B's mission statement and program goals focused on teaching critical thinking, practical application of theoretical understanding, and risk-taking in learning. The conceptual framework of the program cited research literature that supports the efficacy of interactive, constructivist learning in authentic environments, as defined in the mission statement and goals. It included studies of situated cognition, which support the practices endorsed by scholars who promote the teaching of higher order thinking through exploratory, authentic learning. An accreditation document claimed that school-based experiences are evaluated using authentic assessments, including portfolio entries.

Site B also focused on affective factors in learning. The accreditation document stated:

> Faculty in the School of Education endeavor to provide guidance and model the affect and strategies necessary to support students' sense of efficacy. Both the undergraduate and the master's program offer several options through which students can take charge of their own learning. Faculty are committed to supporting students, and they are reviewed – and rewarded – for advising and teaching.

However, implementation of the portfolio process was not as well developed as the philosophical statements, partly because of the lack of clear direction from the state department. Even before securing approval from the state board of education, staff members at the state department of education notified teacher preparation programs of changing accreditation requirements. The state board delayed adopting the new standards, thereby postponing the required transition to a new standards-based model.

The changes in requirements led to a protracted implementation process. Previously faculty at Site B had spent several hours developing program goals for their undergraduate and graduate programs and had just begun discussions of how the goals would be assessed. When the state adapted the NCATE standards, there was not an exact correspondence between Site B's goals and the state's standards. One faculty member volunteered to study where there was agreement between the two documents. She found that Site B's goals met or exceeded all of the standards with one exception; assessment was embedded in the extant goals and needed to be highlighted under the state standards. The professor concluded that meeting this requirement by emphasizing assessment would strengthen the program.

When the state department established its standards, faculty considered ways to modify the existing requirements. For example, the original guides for preparing portfolios instructed students to organize their portfolios around Site B's goals. One faculty member who incorporated portfolios in her course expressed some concern about abandoning Site B's goals because they contained all of the elements of the standards. The program director, however, suggested that formatting the portfolio around the standards would communicate better to state department staff how each student had met the standard.

Because faculty at Site B, by their own admission, had a history of resisting external oversight, the organizing structure of the portfolio consumed a great deal of attention. Meanwhile, the director of field services prepared new student handbooks to reflect the standards, knowing she would need to revise them again when the program itself was better aligned to the standards.

At the time of the study Site B had only two years experience with portfolio assessment of pre-service teachers. They intended to begin by piloting portfolios among student teachers because the state required only documentation that the outcomes had been achieved. One faculty member took responsibility for attending informational meetings held by the state department, conducting portfolio workshops for students, and reading all the student teachers' portfolios. She then met individually with each student and provided oral feedback. Simultaneously the faculty member of the introductory professional course incorporated portfolios at the one PDS site and in her campus-based course. She secured external funding to pay evaluators to compare the portfolios from the two settings.

Such independence of the faculty at Site B was evident in other respects as well. There was no structure in place to facilitate faculty teamwork in development activities. In discussions about the portfolio requirement, some faculty members resisted very vocally, citing the time commitment. They said that they valued personal relationships with students, and this was documented by the number of hours each faculty member met with individual advisees. However, only after several informational sessions did a majority of faculty members view the portfolio as a means of relating to their students.

The brief descriptions of the two sites illustrate that characterizing their quality could challenge a program evaluator. To ascertain how researchers have studied data from portfolios, a literature review was conducted on the use of portfolios in teacher education.

METAPHORS IN PORTFOLIO LITERATURE

In *Metaphors We Live By*, Lakoff and Johnson (1980) stated that metaphor is pervasive in the language, thought, and action of everyday life. They contended that, because humans are not normally aware of their own conceptual systems, looking at metaphors provides evidence of thought that is not available directly. Applying Lakoff and Johnson's notion, the metaphors used to describe portfolios should help elucidate their appropriate use as assessment techniques. Throughout the literature, the metaphors that authors used to describe portfolios were strikingly similar. For this reason, the themes of the metaphors became the organizing structure for the literature review in this section.

Of course, the teaching portfolio is itself a metaphor that educators have borrowed from visual artists. Yet academic or teaching portfolios have taken on their own identity, so the second level of metaphor is the focus here.

Generally, the literature analysis revealed the three metaphors of portfolios as statistics, portfolios as tools, and portfolios as conversations.

Portfolios as Statistics

One of the metaphors to emerge consists of statistical terms generally associated with research literature, such as reliability and validity. Some might argue that statistical references in these works are themes, designs, or procedures rather than metaphors. However, the requisite assumptions for statistical analyses (e.g. Pedhazur, 1997) are frequently lacking from studies of program portfolios. Rather than being statistical studies, these articles appear to use the metaphors of statistics as a means of endowing portfolios with the status of standardized measures. This description reflects Lakoff and Johnson's (1980) definition of metaphor.

The metaphor of statistics is congruent with the usual role of portfolios – student assessment – and can be found in several program descriptions. Fahey and Fingon (1997), for example, describe methods for ensuring interobserver reliability. Nweke and Noland (1996) correlated grades in courses with those on portfolios. Krause (1996) compared two groups of students' understanding of the portfolio process, varying the treatment in each group. Naizer (1997) studied the reliability and validity of the portfolio process by assessing students' domain-strategic and general-learning-strategic knowledge.

There are, of course, statistical studies on the reliability and validity of portfolios that do meet the assumptions necessary for making high-stakes decisions (e.g. Myford & Mislevy, 1995; Resnick, 1996). However, given the normal uses of portfolios in teacher education programs, such statistical rigor is not required and may even divert attention from the important factors that portfolios bring to teacher preparation and the evaluation of quality in those programs.

Portfolios as Tools

A second metaphor found in the literature was the portfolio as tool. This metaphor is congruent with the perspective that portfolios provide appropriate assessments in programs that are based on situated cognition because the theme of tools frequently appears in the literature on situated cognition (e.g. Cognition and Technology Group at Vanderbilt, 1990). Further, in their review of 24 teacher education programs, Anderson and DeMeulle (1998) found that assessment by portfolio was associated with a constructivist model of teacher preparation. In the articles under review here, the portfolio was described as a tool for: (a) *reflection*

(Borko, Michalec, Timmons, & Siddle, 1997; Copenhaver, Waggoner, Young & James, 1997; Holt, Marquardt, Boulware, Bratina & Johnson, 1997; Lyons, 1998); (b) *inquiry* (Grant & Huebner, 1998; Snyder, Lippincott & Bower, 1998); and (c) *professional development* (Freidus, 1996; Wolf & Dietz, 1998).

The metaphor of the portfolio as tool provides important information because it focuses the evaluator as well as the student on the appropriate application of the tool. In fact, Snyder et al. (1998) warned that the portfolio might be a tool that is used for too many or even competing uses. For example, they found that using portfolios for inquiry into personal practice during pre-service teacher professional development and as a means of evaluation might result in a troublesome tension.

Portfolios as Conversations

The third metaphor, portfolio as conversation, was cited specifically by Bartell, Kaye and Morin (1998). They claimed that conversations about portfolio entries offer in-service teachers the best opportunity to grow, learn, and enhance teaching expertise. Interestingly, these authors used the metaphor of a journey in the title of their article whereas the paper itself describes the conversations around the portfolios. Wolf, Whinery and Hagarty (1995) also described portfolios as conversations. Their use of conversation, however, is limited because the authors prescribe how to conduct the sessions "properly," thereby contradicting the notion of a free exchange of ideas that is generally implied in a conversation.

Even studies that did not specify the metaphor of conversation describe activities that may be characterized as such. For example, Cole and Ryan (1998) and Freidus (1996) discussed how the concepts related to the portfolio are communicated through courses, mentoring, and authentic experiences. Berry, Kisch, Ryan and Uphoff (1991) noted the dialogue between students and faculty that occurred as students communicated their growth and faculty responded to students' comments and needs. Loughran and Corrigan (1995) found that only when students had the notion of audience (i.e. prospective employer) did they understand the value of the portfolio. Putting the task in a communicative context helped teacher educators convey the meaning of the portfolio to their students.

The metaphor of the portfolio as conversation emerged as the most potent and persuasive. In the scholarly literature, the conversation appears from the time that students are taught how to prepare the portfolio until they receive feedback that they have achieved the final outcomes. The conversation occurs in writing and speaking, between faculty and students, among faculty and students in peer groups, and as inner speech rehearsal by an individual student. The conversation may be initiated by the state or national accrediting or governance body, faculty, or raters.

The metaphor of the conversation is also evident in the programs at the two sites described in this chapter. Despite the differences in the universities, both programs represented portfolios as conversations. Examples of the conversations at the two sites and in the research literature are elaborated in the next section by comparing linguistic descriptions of conversations with activities surrounding the implementation of portfolios.

DEFINITIONS OF CONVERSATION

Conversation is complex, intentional, and variable. A review of classic linguistic descriptions of conversational strategies and how children acquire them should clarify how the metaphor of conversation may guide evaluators. Some themes found in linguistic descriptions of conversation are synthesized below with the roles of portfolios as conversation found in the literature review and program observations.

Conversation Affects Action

Linguists frequently describe conversations in terms of purposeful social interchange. Pike (1982), for example, noted that language is action, a kind of behavior: "When people talk to other people, they may wish to influence them to act differently, to believe differently, or to interrelate with them in some social way. If language did not affect behavior, it could have no meaning" (p. 15). This definition calls to mind Resnick's (1996) assertion that performance assessment consists of "a set of procedures designed to serve certain social and institutional functions" (p. 3).

The portfolio as conversation in action, especially the behavior designed to influence others, was clear in the literature and practice. Most obviously, pre-service students wanted to persuade faculty members that they should be certified as teachers. Less evident was the use of the portfolio as a means for the state authorities to convince the public that they were leading school reform. For instance, at both sites the state departments required that pre-service teachers document that they had affected children's learning. Perhaps the least obvious intervention was that the portfolio was used in the teacher education program to educate faculty members about authentic teaching and assessment.

Learners Acquire Language and Culture Simultaneously

A second principle of conversation is that the context provides meaning as much as words themselves (Pike, 1982). Although all languages appear to have the same set of illocutionary acts and strategies, they differ significantly in when to perform a speech act and with what strategy (Fraser, 1979). However, children learn these strategies as they learn to talk. Garvey (1975) found that by 5.5 years of age children had learned conversational structure, namely getting attention, taking turns, making relevant utterances, nominating and acknowledging topics, ignoring and avoiding topics, priming topics, and requesting clarification. Later Garvey (1977) demonstrated that learning to talk is learning how to interact.

Applying these conversational strategies to pre-service teachers' portfolios, it becomes clear why students ask questions and incorporate materials into their portfolios as they do. As pre-service teachers learn to use their portfolio, they attempt to get the readers' attention – at times with overly decorative covers or other inappropriate materials. Students struggle to learn what they can say to whom and when they can say it. They aspire to make relevant utterances. They nominate their own topics and acknowledge the topics that readers require in the portfolio, and they ignore or avoid those topics that confuse or annoy them. As they develop proficiency, student teachers become more proficient in accurately communicating their professional growth. Generally, the portfolio provides an important context for learning how to interact professionally.

As in initiating conversations, the portfolio often begins with a question or priming topic. In Site A a question guided students as they prepared to enter each of the three phases of the program: (a) Why do you want to be a teacher? (b) What evidence do you have that you are ready to student teach? and (c) What evidence do you have that you are safe to start teaching?

After the conversation topics are nominated, interlocutors need appropriate information to be able to continue the conversation. At Sites A and B and in the literature, effective portfolios were used in conjunction with authentic experiences that promoted professional interaction.

To maintain a conversation, the complexity of discourse requires that both interlocutors use strategies appropriate to the context. In practice and in the cited works portfolios were used in conjunction with authentic experiences that provided a context for professional social interaction. After the topics were initiated, pre-service teachers could relate their experiences and incorporate evidence of their professional development to maintain the portfolio as conversation.

Language Learning Requires Simplification

Novice portfolio writers, similar to novice language learners, understand that their communicative task is complex. One strategy that learners use is simplification, in which the learner attempts to make the enormous task more manageable by reducing its complexity. For example, a simplification strategy of some students at Sites A and B was to follow a guide strictly and/or to write no or minimal information about each entry. A more effective strategy was to rely on feedback from those with greater knowledge: peers, faculty members, and mentoring teachers.

Like language learners, portfolio writers become more proficient with more practice and feedback, but growth is idiosyncratic and non-linear. Eventually the new language becomes more automatic. One student at Site A remarked how she had begun her portfolio by following all the rules, but by the time she presented her portfolio to the committee, it reflected her own voice. This statement provides evidence of the qualitative difference between the novices and those with growing expertise.

The complexity of the portfolio as conversation is apparent in other contexts as well. As was observed in the two sites, conversations are occurring on many levels at once. Portfolios may reflect conversations in the foreground, background, behind the scenes, and in flashbacks. For example, in the context of state agencies, portfolios might be intended to communicate to the public that the state is making teacher preparation programs accountable for certifying well-prepared teachers. In Site B flashbacks to conversations about external oversight provided a context that the program director had to consider as she attempted to advance the conversation about portfolios among the faculty. Students at both sites carried on peer conversations behind the scenes that both supported and interfered with effective communication.

Conversations Require Feedback

Effective communicators maintain a conversation by employing strategies that are appropriate to the context. Linguists study these strategies by distinguishing the types of social interaction. Perhaps their distinctions can enhance educators' and evaluators' understanding of how the portfolio conversation can be refined.

In his classic work, Joos (1967) clarified that conversations may be classified according to the amount of contextualization or shared knowledge and the amount of feedback. On either end of Joos's taxonomy are intimate conversations, which assume mutually known background information, and formal or frozen language, which requires no prior understanding and allows no feedback.

Generally, portfolios as conversations are at Joos's mid-level, the consultative. At that level, one interlocutor has information that the other does not have. The expert must convey information to the listener or reader with sufficient detail so that, at the end of the conversation, both have similar understandings of the material that was shared. The expert also looks for feedback to be able to change the conversation if the listener does not appear to understand or agree. To be effective, portfolios have to provide a significant number of ideas to meet the consultative-level requirements. That is, the expert must provide detail, yet check for understanding and adjust the information based on feedback from the listener/reader.

The portfolio raters in Myford and Mislevy's (1995) investigation implied that portfolio writers should communicate at the consultative level. They wrote, "students should strive for clarity, coherence, and consistency" (p. 35) when writing their portfolios. Examples of effective uses of portfolios (e.g. Borko et al., 1997; Copenhaver et al., 1997; Grant & Huebner, 1998; Holt et al., 1997; Snyder et al., 1998) also reflected the consultative level as teacher education students documented inquiry or reflection and the consequent changes in their own teaching or children's learning.

Teacher educators also model the consultative level. That is, faculty provide guidelines for the portfolios, adjusting the information according to student feedback. Some programs resorted to the formal level of communication – where the expert provides information without seeking feedback from the listener – when they provide portfolio guidelines and requirements instead of a real conversation. In linguistic terms, not all social interaction is conversation (Pike, 1982).

In effective programs the communication is two-way: Instructions for effective portfolios require dialogue and feedback. For example, Freidus (1996) explained how detailed published guidelines were used in conjunction with ongoing meetings between faculty mentors and students and among students themselves for peer mentoring. Both Sites A and B had introductory workshops, opportunities for peer mentoring, and faculty mentors to clarify and support the portfolio as conversation. In addition, Snow's (this volume) recommendations for communicating quality suggest an interactive process that faculty could adapt to encourage students to reflect on how they are conveying quality in their portfolios.

CONCLUSION

Portfolios provide a unique opportunity for teacher educators to communicate and represent the quality of their programs. Portfolios provide a variety of longitudinal data to document development (Touzel, 1993). Program evaluators can judge not only the amount and kinds of skills and dispositions the learners acquired but also the extent to which the participants' learning reflects the programs' goals. Those goals provide the common framework of meaning (Myford & Mislevy, 1995) that informs the stakeholders, whereas the portfolios reflect each learner's ability to construct meaning around their individual learning.

These characteristics provide exceptional advantages to evaluators, including program-accrediting teams. Rather than using traditional forms of evaluation, program-accrediting teams could evaluate teacher education programs holistically. As some of the authors of this book have stated, evaluations are useful if they provide fresh perspectives on the quality of the evaluand. To achieve this, Stake (this volume) urges evaluators to understand quality in terms of the experiences of the stakeholders. Identifying metaphors that reflect participants' conceptual systems (Lakoff & Johnson, 1980) related to the evaluand may be a useful approach to finding those fresh perspectives.

For example, the elements of conversation found in portfolios might provide a framework by which the evaluator gains insights into the learners' experiences in a program. Some example questions based on the conversation metaphor include the following: How did the learners communicate that they constructed knowledge, reflected on experiences, and solved problems? Do the portfolios demonstrate greater sophistication over time? Did the portfolio entries compel the readers to act on any information? How do the writers communicate the quality of the characteristics of their program?

Evaluators searching for program quality can also make judgments based on metaphors. Continuing with the conversation metaphor, evaluators may ask questions such as the following: Do the portfolios provide evidence that faculty engaged in dialogues with the learners? Did the faculty learn about the experiences in the program through the portfolios and act to modify ineffective practices? What critical topics do the portfolio writers avoid? How do the portfolios communicate the quality of the program? The challenge for teacher preparation programs is to find ways to promote professional growth through conversations that are succinct and scintillating. Program evaluation becomes another element in the conversation, providing feedback to the program participants and encouraging them to find new ways to represent quality through the voices of their stakeholders.

The current politicization of the evaluation of school programs challenges evaluators to consider carefully the issues surrounding the evaluation of quality that are raised in this book. The stakes are high for the participants. Teacher education students who have invested at least four years meeting input requirements will probably be judged on the basis of output measures that may or may not reflect the students' teaching abilities. Programs that have invested capital and human resources to meet state requirements may lose accreditation if program evaluators find the graduates' skills lacking.

But, because the stakes are high, educational evaluators have the opportunity to uncover quality. They need to consider whether traditional evaluation paradigms, such as analyzing numbers from standardized tests, are providing an accurate representation of quality. They can utilize techniques not necessarily included in traditional evaluation paradigms, experiment with creative approaches to collecting and analyzing data – such as describing quality through metaphor – and enrich the experiences of all the stakeholders as they participate in evaluation and regeneration processes.

REFERENCES

Anderson, R. S., & DeMeulle, L. (1998). Portfolio use in twenty–four teacher education programs. *Teacher Education Quarterly*, *25*(1), 23–31.

Bartell, C. A., Kaye, C., & Morin, J. A. (1998). Portfolio conversation: A mentored journey. *Teacher Education Quarterly*, *25*(1), 129–139.

Berry, D. M., Kisch, J. A., Ryan, C. W., & Uphoff, J. K. (1991, April). *The process and product of portfolio construction.* Paper presented at the Annual Meeting of the American Educational Research Association, Chicago, IL. (ERIC Document Reproduction Service No. ED 332 995)

Borko, H., Michalec, P., Timmons, M., & Siddle, J. (1997). Student teaching portfolios: A tool for promoting reflective practice. *Journal of Teacher Education*, *48*(5), 345–357.

Brown, J. S., Collins, A., & Duguid, P. (1989). Situated cognition and the culture of learning. *Educational Researcher*, *18*(1), 32–42.

Cognition and Technology Group at Vanderbilt (1990). Anchored instruction and its relationship to situated cognition. *Educational Researcher*, *18*(1), 2–10.

Cole, D. J., & Ryan, C. W. (1998, February). *Documentation of teacher education field experiences of professional year interns via electronic portfolios.* Paper presented at the Annual Meeting of the Association of Teacher Education, Dallas, TX. (ERIC Document Reproduction Service No. ED 418 057)

Copenhaver, R., Waggoner, J. E., Young, A., & James, T. L. (1997). Promoting preservice teachers' professional growth through developmental portfolios. *Teacher Educator*, *33*(2), 103–111.

Fahey, P. A., & Fingon, J. C. (1997). Assessing oral presentations of student-teacher showcase portfolios. *Educational Forum*, *61*(4), 354–359.

Fraser, B. (1979, February). *Research in pragmatics in second language acquisition: The state of the art.* Paper presented at the annual meeting of Teachers of English to Speakers of Other Languages, Boston.

Freidus, H. (1996, April). Portfolios: *A pedagogy of possibility in teacher education.* Paper presented at the Annual Meeting of the American Educational Research Association, New York. (ERIC Document Reproduction Service No. ED 395 915)

Garvey, C. (1975). Requests and responses in children's speech. *Journal of Child Language*, *2*, 41–63.

Garvey, C. (1977). The contingent query: A dependent act in conversation. In: L. M. Rosenblum & L. A. Rosenblum (Eds), *Interaction, Conversation and the Development of Language* (pp. 63–93). New York: John Wiley.

Grant, G. E., & Huebner, T. A. (1998). The portfolio question: A powerful synthesis of the personal and professional. *Teacher Education Quarterly*, *25*(1), 33–43.

Holt, D., Marquardt, F. M., Boulware, A., Bratina, T., & Johnson, A. C. (1997, February). *Integrating preparation and practice through a technology-based approach to portfolios for professional development using CD-ROM technology*. Paper presented at the Annual Meeting of the American Association of Colleges for Teacher Education, Phoenix, AZ. (ERIC Document Reproduction Service No. ED 405 324)

Joos, M. (1967). *The five clocks.* New York: Harcourt Brace and World.

Krause, S. (1996). Portfolios in teacher education: Effects of instruction on preservice teachers' early comprehension of the portfolio process. *Journal of Teacher Education, 47*(2), 130–138.

Lakoff, G., & Johnson, M. (1980). *Metaphors we live by.* Chicago: University of Chicago.

Loughran, J., & Corrigan, D. (1995). Teaching portfolios: A strategy for developing learning and teaching in preservice education. *Teaching and Teacher Education, 11*(6), 565–577.

Lyons, N. (1998). Reflection in teaching: Can it be developmental? A portfolio perspective. *Teacher Education Quarterly, 25*(1), 115–127.

Myford, C. M., & Mislevy, R. J. (1995). *Monitoring and improving a portfolio assessment system* (CSE Technical Report No. 402). Los Angeles: University of California, National Center for Research on Evaluation, Standards, and Student Testing.

Naizer, G. L. (1997). Validity and reliability issues of performance-portfolio assessment. *Action in Teacher Education, 18*(4), 1–9.

Nweke, W. C. (1991). *What type of evidence is provided through the portfolio assessment method?* Paper presented at the Annual Meeting of the Mid-South Educational Research Association, Lexington, KY. (ERIC Document Reproduction Service No. ED 340 719)

Nweke, W., & Noland, J. (1996, February). *Diversity in teacher assessment: What's working, what's not?* Paper presented at the Annual Meeting of the American Association of Colleges for Teacher Education, Chicago, IL. (ERIC Document Reproduction Service No. ED 393 828)

Pedhazur, E. J. (1997). *Multiple regression in behavioral research: Explanation and prediction.* (3rd ed.). Ft. Worth, TX: Harcourt Brace College.

Pike, K. L. (1982). *Linguistic concepts: An introduction to tagmemics.* Lincoln, NE: University of Nebraska Press.

Resnick, L. (1996). *Performance puzzles: Issues in measuring capabilities and certifying accomplishments.* (CSE Technical Report No. 415). Los Angeles: University of California, National Center for Research on Evaluation, Standards, and Student Testing.

Shavelson, R. J., Lang, H., & Lewin, B. (1994). *On concept maps as potential "authentic" assessments in science.* (CSE Technical Report 388). Los Angeles: University of California, National Center for Research on Evaluation, Standards, and Student Testing.

Snyder, J., Lippincott, A., & Bower, D. (1998). The inherent tensions in the multiple uses of portfolios in teacher education. *Teacher Education Quarterly, 25*(1), 45–60.

Tharp, R. G., & Gallimore, R. (1988). *Rousing minds to life: Teaching, learning, and schooling in social context.* New York: Cambridge University Press.

Touzel, T. J. (1993, February). *Portfolio analysis: Windows of competence.* Paper presented at the Annual Meeting of the American Association of Colleges for Teacher Education, San Diego, CA. (ERIC Document Reproduction Service No. ED 356 207)

Wolf, K., Whinery, B., & Hagerty, P. (1995). Teaching portfolios and portfolio conversations for teacher educators and teachers. *Action in Teacher Education, 17*(1), 30–39.

Wolf, K., & Dietz, M. (1998). Teaching portfolios: Purposes and possibilities. *Teacher Education Quarterly, 25*(1), 9–22.

16. PERFORMANCE-BASED ASSESSMENT: A QUALITY IMPROVEMENT STRATEGY

Delwyn L. Harnisch

ABSTRACT

This paper presents a performance-based view of assessment design with a focus on why we need to build tests that are mindful of standards and content outlines and on what such standards and outlines require. Designing assessment for meaningful educational feedback is a difficult task. The assessment designer must meet the requirements of content standards, the standards for evaluation instrument design, and the societal and institutional expectations of schooling. At the same time, the designer must create challenges that are intellectually interesting and educationally valuable. To improve student assessment, we need to design standards that are not only clearer, but also backed by a more explicit review system. To meet the whole range of student needs and to begin fulfilling the educational purposes of assessment, we need to rethink not only the way we design, but also the way we supervise the process, usage, and reporting of assessments. This paper outlines how assessment design is parallel to student performance and illustrates how this is accomplished through intelligent trial and error, using feedback to make incremental progress toward design standards.

Visions of Quality, Volume 7, pages 253–272.

ISBN: 0-7623-0771-4

INTRODUCTION

Why should educators pay close attention to the way standardized tests are designed? All too often, mandated assessments that require huge expenditures of time and energy are intended to serve purposes external to the school and have little to do with teaching and learning in the classroom. In a system of schooling that relies on externally developed policies and mandates to assure public accountability, teachers are held accountable for implementing curriculum and testing policies that are prescribed by governmental agencies far removed from the school setting (Darling-Hammond, 1988). What then is the stake in this endeavor for teachers and students? As their cooperation is required, their participation should be aimed at providing them with clear and tangible benefits. If the improvement of teaching and learning plays a significant role in their design, assessments can more directly be used to support specific advances in instruction and student performance.

Designing assessment is like practicing architecture. The architect must work not only with opportunities, but also within given constraints such as the functional needs and local aesthetics that inform building codes and shape specific design values. In creating a test for purposes of student assessment, the test designer must meet the requirements of content standards, standards for the design of evaluation instruments, and societal and institutional expectations of schooling. At the same time, the test designer must make test questions intellectually challenging and of high value to clients.

The typical situation in assessment design, however, may be viewed as the reverse of architecture. The primary aim in architecture is to please the client, while carefully adhering to an understanding of the necessary constraints. In school test design, we often see the designers worrying about satisfying their own needs (efficient administration and easy-to-score tests) or construction codes (technical test and measurement standards) instead of serving students' interest. To produce assessments of educational value and of high quality, designers must follow students' needs for more challenging and useful work and teachers' requests for more direct, timely, useful, and practical information.

Purpose

The purpose of this chapter is to outline a set of test design standards and to discuss the possible ways that these standards can be implemented at the school level. In architectural design we see that decisions about content, scale, and aesthetics are discussed in terms of function or purpose of space. Consider, for example, the following questions on building a new school: What will be the best building design for the classrooms? What kind of feel, people flow, and scale are appropriate for the target age group of students? What kind of design best fits with the purposes of the schools and its constituents? Decisions about assessment design would also benefit from considering form and purpose.

The old model from architecture, "Form follows function," might well apply to assessment. Many of us have not attended to the foremost necessity of using assessments to help us better understand students' needs. We have instead focused on accountability needs and used old test forms. In the process of utilizing old test forms, which are easy to score and secure, we have lost focus on our obligations to students.

To improve student assessment, we need to have design standards that are clear and more direct in helping us to understand our students and to review our existing assessments. It is time to stop ignoring the range of student needs and to begin providing meaningful educational feedback to our students. In this process we need to rethink both the way we design assessments and the manner in which we provide educational feedback through the use and reporting of assessments.

In performance-based testing, we need to build tests that are consistent with standards and content outlines. In the process of test construction, we will show how the process of test design parallels student performance. Tests may be improved through an intelligent trial and error process, using feedback to make progress toward design standards.

Defining Standards

A standard points to and describes a desirable level of exemplary performance. A standard is a worthwhile target, irrespective of whether or not most students can or cannot meet it at the moment. Standards are made concrete by criteria. The expectation that students should leave school able to communicate well is an example of a performance standard for which we may establish criteria to identify what we mean by well. We can thus describe students as communicating well when they are communicating with clarity, purpose, and understanding of audience. Such criteria are necessary, but not sufficient. To set a standard using such criteria, we need to show just how clear and engaging the communication has to be by employing examples, models, or specifications.

The 1993 report to the National Education Goals panel, *Promises to Keep: Creating High Standards for American Students*, notes a distinction between content standards and performance standards. The report specifies that content standards mean "what students should know and be able to do" (p. 9) and that performance standards specify "how good is good enough" (p. 22). A typical performance standard indicates the nature of the evidence (i.e. an essay, mathematical proof, scientific experiment, project, exam, or any combination of these) required to demonstrate that the content standard has been met and that the quality of student performance is deemed acceptable. Quality, in this context, refers to the level of student performance on a scale in which excellence or perfect achievement is the highest order of accomplishment.

Criteria for Performance Assessments

Criteria differ from standards in other ways. A sports example might be the high jump. To be successful, high jumpers must clear the bar. The question is at what height should we place the bar? This is a performance standard question. The need for standards also explains why a rubric is necessary, but not sufficient. Descriptors in rubrics often refer to generalized criteria. The particular height of the bar in a high jump is an example of setting a performance standard; thus specifying at what level the criteria must be met.

Given this understanding of assessment, the key concerns on which we must focus are the kinds of assessment and the criteria for assessment design that would help make the testing process more educational as well as more technically adequate. Improved assessment may serve not only to set more effectively the bar of performance expectations, but also to measure and represent more accurately the quality of that performance.

What design standards of the performance task best honor the student performance standards being discussed at state and national boards? The standards presented in the recent document by the National Council of Teachers of Mathematics (1998) are worthy and informative targets for the math community. In some countries, however, these standards might strike the readers as unrealistic. The teachers of mathematics in Japan recently discussed these standards and looked at the match with the national syllabus in mathematics. Some early comments from these discussions reveal that the Japanese teachers view these standards as unrealistic. This perspective may represent a misunderstanding of what a standard is meant to be, specifically that a standard embodies the principles that serve as the model against which all judgments about current practices are made. What performance standards provide is an umbrella or a self-correcting system that consists of a mix of reality and feedback to move the system from what is toward what ought to be (CLASS, 1993).

The Japanese concept of Kaizen suggests that quality is achieved through constant, incremental improvement. According to W. Edwards Deming (1986), guru of the Total Quality Management movement, quality in manufacturing is not achieved through end-of-line inspections; by then, it is too late. Rather, quality is the result of regular inspections (assessments) along the way, followed by needed adjustments based on the information gleaned from the inspections. What is the relationship of these ideas to the classroom setting? We know that students will rarely perform at high levels on challenging learning tasks on the first attempt. Deep understanding or high levels of proficiencies are achieved only as a result of trial, practice, adjustments based on feedback, and more practice. Performance-based instruction underscores the importance of using assessments to provide information to guide improvement through the learning process, instead of waiting to give feedback at the end of instruction. Kaizen, in the context of schools, means ensuring that assessment enhances performance, not simply measures it.

Three Different Kinds of Educational Standards

Before discussing the design standards for assessment, it is important to understand the role that they play in the specific performance standards for students and in the larger realm of educational standards. The three essential standards for designing assessments are: (a) *Content standards* that answer the question of what students should know and be able to do; (b) *Performance standards* that answer the question of how well students must do their work; and (c) *Task or work design standards* that answer the questions of what constitutes worthy and rigorous work and what tasks students should be able to do.

Each of these standards is needed in developing high quality assessments. It is important that we state performance standards in specific measurable terms to assure that the challenge to content is not subverted by low standards and expectations. Most students want to know what counts as mathematical problem solving and what standard will be used to judge whether the solution is adequate or not. Having standards that are understood in measurable terms allows teachers to aim their teaching toward these specific standards.

The following example may help to illustrate the relationship of these standards. In *Can Students Do Mathematical Problem Solving*? (Dossey, Mullis & Jones, 1993), the National Assessment of Educational Progress (NAEP) provide the following performance task as a measure of mathematical competence:

> Treena won a 7-day scholarship worth $1,000 to the Pro Shot Basketball Camp. Round-trip travel expenses to the camp are $335 by air or $125 by train. At the camp she must choose between a week of individual instruction at $60 per day and a week of group instruction at $40 per day. Treena's food and other expenses are fixed at $45 per day. If she does not plan to spend any money other than the scholarship, what are all choices of travel and instruction plans that she could afford to make? Explain your reasoning (p. 116).

This task was accompanied by a five-point rubric that describes various levels of performance. One of these levels is identified as the performance standard, the acceptable level of performance on the task. In the case of the NAEP performance task, level four on the rubric is considered the performance standard. It states, "The student shows correct mathematical evidence that Treena has three options, but the supporting work is incomplete. Or: The student shows correct mathematical evidence for any two of Treena's three options, and the supporting work is clear and complete" (p. 119).

This example reveals that setting standards depends on two decisions about what we mean by an acceptable performance. Standards must consist of a combination of a meaningful task and a quality of performance. For schools to improve, content and performance standard guidelines for what counts as important kinds of work at each grade level are needed, along with an assessment system that sends the message that such work is mandatory.

An example from athletics may further demonstrate the need for an assessment standard. The decathlon was invented about 100 years ago for the Olympics. Many people around the globe claimed to be great all-around athletes, but there was no agreement on what that claim meant or how to measure it. A small group of people decided that great all-around athletes were those who could qualify to compete in a decathlon, ten specific tasks that, according to the Olympic founders, tapped the full range of genuine track and field performance. This compares, perhaps, to where we are in several of our disciplines in education in which there is little agreement on what constitutes required work worth mastering, despite the fact that we have clear guidelines for what schools should teach.

What is needed at the school level is agreement on the performance tasks that would count as the basis for evidence of general intellectual excellence within and across subjects. These tasks would be limited to a manageable sample of complex tasks from the whole domain of intellectual performance. These sorts of performance genres would be examined with respect to the content standards for each subject. Examples of these oral performance genres include a speech, report, proposal, debate, discussion, or simulation. Written examples include a letter, critique, script, plan, poem, log, narrative, description, or an essay or analysis. Visual display examples include an exhibit, advertisement, artistic medium, electronic media; or a model, blueprint, graph, chart, or table.

World Class Standards

Having a self-correcting system works best when there is clear agreement about standards. We have world-class records in athletic and artistic arenas. This is a primary reason why athletic and artistic performances continue to improve over time, while educational performances do not. Coaches plan their strategies to improve their athletes' standing by having them engage in tasks intended to improve their performance standards. Architects know that to be successful in their profession they need to continue to improve building designs to satisfy current building codes and their clients' needs. The challenge for education is to agree on the standards for tasks and performance and to use those standards and tasks to assist in the teaching and learning process. An additional challenge is to establish a process for implementing activities that will improve teachers' design of assessments as well as student performance.

Challenges Facing Performance Standards

One of the biggest challenges facing the use of performance standards is their specificity or lack of generalizability. Shavelson and his colleagues (Shavelson, Baxter & Pine, 1992; Shavelson, Baxter & Gao, 1993) conducted a series of studies that investigated the uses of performance tests. They discovered through their research that a student's performance on a single task is not necessarily a good indicator of the student's knowledge and skill within the subject area that the task is designed to represent. Shavelson and his colleagues tested this by giving students the same science task in three different formats: hands on, computer simulated, and written descriptions derived from a hands-on experiment. They found that students might perform well in one format, but not in the other two. They concluded that a single performance assessment is not a good general indicator of how well students can perform within a content area. In fact, most measurement experts now contend that anywhere from 5 to 23 performance tasks are necessary to assess accurately students' competence within a single subject area, depending on the specific study (Shavelson et al., 1992).

Some schools are setting performance standards on traditional achievement tests. This process, called "cut scores", is commonly used by experts to make judgments about which items students who have a basic understanding of the subject area being assessed should be able to answer correctly. If we take a traditional test and apply a performance standard that uses a total number of points representing correct responses on specific items, we can see that this score means that students must answer items on a variety of topics. The following is an example of the topics required for a proficient level of performance, as defined by *Setting Performance Standards for Student Achievement* (NAEP, 1993):

> Fourth graders performing at the proficient level should be able to use whole numbers to estimate, compute, and determine whether results are reasonable. They should have a conceptual understanding of fractions and decimals; be able to solve real world problems in all NAEP content areas; and use four-function calculators, rulers, and geometric shapes appropriately. Students performing at the proficient level should employ problem-solving strategies, such as identifying and using appropriate information. Their written solutions should be organized and presented both with supporting information and explanations and how they were achieved (p. 34).

EVIDENCE OF QUALITY

The rationale for performance testing rests largely on its consequences for teaching and learning. In general, we value assessments because they give us accurate information for planning, decision-making, grading, and other purposes. They also provide a measure of fairness in making judgments and actions.

If our performance tasks lack sufficient technical quality, the assessment results will provide misinformation about individual students, classrooms, schools, and districts. In preparing quality assessments, we must be able to demonstrate evidence of validity and reliability, including the degree to which intended uses of assessments are justified and the degree to which the scores are free of measurement error.

Reliability and Validity

When we consider to what extent performance assessments are reliable, we should remember that we expect the measure of our height to be the same regardless of whose yardstick is used. We also want our measurements of student performance to be reliable and consistent. As performance assessments typically require that scorers or raters evaluate students' responses, they pose special reliability challenges. Raters judging student performance must be in basic agreement, within some reasonable limits, about what scores should be assigned to students' work. If teachers or raters are not in agreement, student scores are measures of who does the scoring rather than the quality of the work. This is not only a technical psychometric issue, but also one that has important implications for school reform.

Some of the factors for teachers not agreeing on student performance scoring suggest that no consensus exists on the meaning of good performance or on the meaning of the underlying standard, just as no agreement exists on expectations for students. Research suggests that the technical process of reaching high levels of rater agreement requires good training and scoring procedures. These include well-documented scoring rubrics exemplified by benchmark or anchor statements, ample opportunities for teachers (scorers) to discuss and practice applying the rubric to student response, and rater checks before and during the scoring process to ensure that evaluators are consistent. Retraining of raters should be used when necessary.

Scoring Rubrics Defined and Utilized

Scoring rubrics communicate what is expected in student performance. The time demands of performance assessments along with the concern for consistent scoring pose some technical problems and challenges. It is these same characteristics of performance assessments that make them valuable to the educational community and to standards-based reform. Performance assessments communicate what is important to learn, while they also provide the teacher with a model of the kinds of instructional tasks and processes that teachers should use in their classrooms.

Research has shown the positive effects of involving teachers in the development and scoring of assessments. Teachers often better understand and come to consensus about what is expected, gain familiarity with teaching practices that are standards-based, and use standard-based ways to assess student understanding. In short, the effects of performance assessments on curriculum and teaching and the value of scoring and scoring rubrics are well-documented (Baker, 1993; Brandt, 1992; Harnisch, 1994, 1995; Harnisch & Mabry, 1993; Herman, 1998; Wiggins, 1998).

Quality Assessment Standards

The process of developing standards for a local school district is a challenging task. The following criteria are helpful in examining assessment practices and in safeguarding the core assumption that assessment should improve performance, not just measure it (Wiggins, 1998).

1. Credible to interested stakeholders, including teachers, students, and parents.
2. Useful both to students and teachers.
3. Balanced in use of multiple assessment methods - providing a profile of achievement.
4. Honest and fair - student performance rated against important standards.
5. Intellectually rigorous and thought provoking - focused on primary ideas, questions and problems.
6. Feasible in terms of resources, politics, and reallocation of time for effectively reporting student progress.

Fig. 1. Assessment Standards

Credibility

A primary goal for every local assessment system should be to have a set of credible assessments. A credible assessment is one that provides data that we can trust and data by which we are willing to be held accountable. Credibility in assessment requires known, agreed upon, and uniform scoring criteria matched with standards. In most local school districts a level of standardization is essential in the administration of performance tasks, criteria, and scoring. For these types of assessments to improve student performance, we need to provide answers to the student and the student's teachers. Summarization of the results should also be credible to parents, board members, and the local community that employs our students. In short, the use of assessments based on clear and concise standards to improve students' intellectual performance provides the means to meet the needs of students and other educational stakeholders.

Usefulness

Assessment methods should be useful to students. This point sounds simple, but it is seldom considered in the design and development process. The assessment design community needs to seek out teacher and student feedback about the quality of the tasks and feedback information. Harnisch, Tatsuoka, and Wilkens (1996) surveyed teachers and students about the quality of feedback from the new score reports for the SAT quantitative reasoning test. Key survey questions addressed the extent to which the reported information was fair and helpful in terms of how the results were reported and analyzed. Usefulness of tests implies not only meaningful reporting of results, but also the opportunity to use them.

Balanced Methods – Anchored in Performance Tasks

Assessment practices should always be focused on measuring effective student performance in terms of skills, learning, and understanding, rather than on simple factual recall exercises. What is needed is a balance of methods that allows our students to achieve good performance on a complex task. In getting students ready to play a soccer game, they must learn how to do various exercises and drills that aid in the development of skills for playing two-on-one soccer. For our students to be able to achieve a good performance on a complex task, we need to model such tasks in their development. The suggestion on balanced assessment is that we evaluate performance using a variety of modes and a sizable number of pieces of evidence. In short, a portfolio can serve as an ideal vehicle for sharing a practical way of honoring the need for the balance in the evidence collected, and at the same time it can anchor the assessment effort in genuine performance tasks and standards.

Honest and Fair – Student Performance Rated Against Important Standards
In working with a student-centered assessment system it is important to have a system that is both honest and fair. Designing assessments for measuring excellence requires that we be honest in that we accurately describe to students their levels of achievement against standards of excellence. We also need to be fair and consider prior experiences, recent changes in academic growth, and any extenuating circumstances. We need to be open and frank with our students, so that we do not cause those who are well below average to think that they are working at an acceptable level.

Intellectually Rigorous and Thought Provoking
The focus of this standard is on producing items that provide students with a challenging task, rather than items of a simplistic nature. Students find challenging problems intrinsically interesting and motivational in nature once they are able to understand them. It is important for teachers to discuss what counts as quality work and to create exercises and tests to measure that level of quality. Watching students work with challenging problems in a collaborative manner shows greater focus than watching students work with meaningless quizzes and worksheets.

Feasibility
This standard focuses on making a good idea happen in the educational system. One of the greatest obstacles to many great ideas in school reform is the lack of time to implement the assessment system properly. A strategy is needed that allows teachers to apply new ways to think about how to use time in school and how to build effective student assessment time into their schedules. This can be done by finding creative and cost-effective ways of providing teachers time to assist one another in performance assessment, designing, and matching performance items and tasks with curriculum goals.

This feasibility standard also involves issues related to professional development and the ways in which we organize our work places. A growing need exists for schools to create an entirely new conception of professional growth, one that makes high quality assessment mandatory and of great importance. To do this requires that we create new descriptions of work for teachers and administrators with a focus on a shared vision and a set of standards that endorses the quality work needed for excellence in assessment.

Using technology can be one means of sharing our models of assessment and ways of providing feedback to our assessment community. School improvement teams can search the Internet and access the databases of approved tasks and rubrics that fit their evaluation systems. Sharing resources approved by school improvement teams on the Internet may be one means of making choices more efficiently available.

Development of Assessment Design

The design of assessments requires careful attention to standards and to the feedback that occurs when we see how our design decisions play out. Assessment design involves knowing the process of self-assessing and making adjustments as the design emerges. It also involves knowing how well or how poorly the design meets local needs. John Dewey (1938) noted that logic is a method of testing and communicating results, rather than a prescription for arriving at good decisions or solutions to problems. Our thinking is consistent with this logic when we employ multiple methods, each providing clear indications of student performance.

Knowing the questions that we want to answer are important starting points. Specifically, this comes by knowing what indicators, criteria, or evidence will be used to qualify student performance on levels from excellent to average to poor. Rules or guidelines are needed to make these discriminations among student performances. Much of what we do in building assessments can be identified as key elements in this process. Omitting various parts of this process opens the door for misunderstanding in the design of assessment. A set of elements consistent with building assessments is identified in Figure 2. Using these elements to review your local assessment design can aid in the process used in building assessments.

Achievement targets or benchmarks may be paired with individual performance tasks in a grade level and subject area (see Appendix for a sample of benchmarks for eighth grade language arts). Additional steps and development are needed, but having a clear target benchmark along with specified performance task aids in the design of the remaining elements.

1. Achievement target described.
2. Performance standards established.
3. Criteria and indicators implied in score report.
4. Implied content standards, different contexts and performance genres.
5. Criteria and indicators implied in content, contexts, and genres.
6. Individual performance tasks designed.
7. Criteria defined for performance tasks.
8. Rubrics constructed.

Fig. 2. Key Elements in Designing Assessments

Effects on Curriculum and Teaching

The effects of performance assessments on curriculum and teaching and the value of scoring and scoring rubrics are well documented in assessment literature. Performance assessments do indeed influence classroom instruction. For example, state standard-based goals like those in Kentucky found an increased coverage of topics measured by the state assessment system and increased use of standards-based teaching and learning practices (Izumi, 1999; Steffy, 1995). Pedagogy changed to include practices such as applying mathematics to real-world problems and requiring students to explain their work by introducing unique and non-routine problems.

Impacts on Thinking about Teaching

Performance assessment activities engage teachers in new ways of thinking about teaching. As teachers engage in the scoring and review of student responses to performance assessments, the process can be a powerful tool for professional development because it can open new insights into student performance. These insights, in turn, can generate new ideas for classroom activities, identify potential gaps in the classroom curriculum, and result in new understanding about students' strengths and weaknesses (Harnisch, Tatsuoka & Wilkens, 1996). Teachers often recognize this process as a redirection of their grading effort toward looking and listening to what students have to say versus simply scoring answers on a page of computation problems.

Educational Value of Rubrics

Rubrics have a powerful effect on teachers. The development and use of rubrics for many teachers gives them a tangible tool to evaluate student work. It also provides a way of thinking about what they are trying to accomplish and a focus for their planning and teaching. Teachers often report how the rubrics have clarified what they wanted to accomplish in the classroom and have given them more time to plan structured lessons that clearly help students to succeed in areas where they knew what was expected. Teachers need to have access to rubrics to assist them in having ways of looking at methods for evaluating student understanding, diagnosing strengths and weaknesses, and providing meaningful feedback to students.

Barriers to Implementation

Time and costs are the two biggest obstacles facing the implementation of standards-based reform. Time is the primary item that needs attention. Teachers need time to become familiar with new assessments, their administration, and the process by which the tasks are developed and scored. They also need time for reflection on results to modify their instructional and assessment practices. More time is needed to implement performance assessments, and time elements represent costs that have important payoffs in classroom practice. Strong administrative support is essential for standards-based reform to succeed.

Costs for implementation vary according to the type of assessment. In my review of performance assessments (Harnisch, 1994), I identified a series of key questions regarding costs: (a) Is someone willing to pay $10 to have a student notebook in science scored and are the measurement properties likely to be achieved? (b) Are the benefits of performance assessments sufficiently greater to justify the increased costs over traditional measures? (c) Are the reporting formats understandable to the student and the teachers to justify the costs? (d) Are sampling strategies appropriate for use in large-scale performance measurement contexts to reduce student costs and yield program or school-level reports?

Improving Student Learning

Performance assessments have been used to measure changes in student learning. Researchers have focused their efforts on understanding whether the use of performance-based assessments affects student learning. Results from the research community are not in agreement. Some say that where performance assessments have been introduced, student scores are initially very low but rise over time, suggesting that student learning has increased. Others say that the gains in achievement are not found on other external measures, such as the NAEP or on other statewide test results. This means that improvement in performance assessment scores does not necessarily indicate that results on other tests will also improve.

Recent findings show that teachers and principals who take these new assessments seriously try to incorporate new pedagogical practices into their classrooms. Many classroom teachers try to engage their students in the kinds of activities that are found on new performance assessments. The interesting question to examine is the extent to which these sustained professional development programs continue to have an impact on student learning (Harnisch, 1997).

CONCLUSION

This chapter has shown the need for greater clarity in the design of assessments. It has also shown how performance assessments alone cannot solve our educational problems, although they can be a beginning step in rethinking the way we measure what children have learned. The process of thinking about these changes allows more teachers to consider the quality of their teaching practices, which is what is needed for more students to achieve the high standards we hold for each of them. When teachers have the opportunity to reflect on the nature and outcome of student performance assessment, they have more immediate access to new ideas and approaches to teaching. They also have a more focused view of what they need to accomplish in the classroom. When assessment clearly demonstrates discrepancies between curriculum expectations and student achievements, teachers can readily identify the areas of greatest need and adjust the content and methodology of their work accordingly. As teachers act on what performance assessments have shown them, students are afforded more opportunities to fill in the gaps discovered and to maintain steady academic progress. In short a clear picture of student understanding and progress enables teachers to offer direct and meaningful assistance where it is most needed. On the other hand, when assessments offer vague, ambiguous perspectives on student learning, they are of little or no value and may serve only as a distraction from productive efforts at teaching and learning.

A goal for our schools is to have more students graduate from high school and to have all those that graduate meet high academic standards. What is needed is improved student learning. What this chapter has outlined are steps in the process that affect or produce learning results and over which schools and teachers can exercise control. Improvement requires control, just as control requires understanding. Control of time is a major concern. Without the ability to set aside time for the study of assessments and assessment results, teachers cannot adequately utilize them as a basis for modifying and improving instruction. Without the opportunity to make full use of assessment to support school improvement, standards-based reform has less chance of success.

Teachers must have a central role in determining curriculum standards if those standards are to have any real meaning in the classroom. Policymakers at each level must recognize their responsibilities reach far beyond merely setting expectations and demanding accountability. Reciprocity between policymakers who determine curriculum standards and classrooms where those standards are to be applied is essential to solving the problems of school settings and achieving the goals of public education. The quick fix of unilaterally raising standards and imposing new accountability systems to deal with perceived shortcomings in education still holds the greatest appeal for legislators and other policymakers who have no immediate ties to the reality of the public school.

If we expect students to improve their performance on these new, more authentic measures, we need to engage in performance-based instruction on a regular basis. A performance-based orientation requires that we think about curriculum not simply as content to be covered but in terms of desired performances of understanding. Thus, performance-oriented teachers consider assessment up front by conceptualizing their learning goals and objectives as performance applications calling for students to demonstrate their understanding. Performance assessments become targets for teaching and learning, as well as serving as a source of evidence that students understand, and are able to apply, what has been taught.

Having established clear performance targets is important for multiple reasons. First, teachers who establish and communicate clear performance targets to their students reflect what is known about effective teaching, which supports the importance of instructional clarity. These teachers also recognize that students' attitudes toward learning are influenced by the degree to which they understand what is expected of them and what the rationale is for various instructional activities. Lastly, the process of establishing performance targets helps identify curriculum priorities, enabling us to focus on the essential and enduring knowledge in our respective field.

To accomplish standard-based reform, standards are needed that effectively focus on student performances. Performance assessments alone will not solve our educational problems, but further thought and reflection on the improvement of performance-based assessment are major factors in advancing the learning that is needed to enable more students to achieve the high standards we expect of them.

ACKNOWLEDGMENTS

The author thanks Art Lehr, Yusaku Otsuka and Shigeru Narita for their helpful comments on various versions of the manuscript. An earlier version of this paper was presented at the 1999 International Conference on Teacher Education held at the Hong Kong Institute of Education. This research was supported while the author was approved by the Japanese Ministry of Education, Science and Culture as a 1998-1999 Research Fellow of the "Outstanding Research Fellow Program" while at Hyogo University of Teacher Education and at the National Institute of Multimedia Education, Makuhari, Japan.

REFERENCES

Baker, E. L. (1993). Questioning the Technical Quality of Performance Assessment. *The School Administrator*, *50*(11), 12–16.

Brandt, R. (Ed.) (1992). *Performance assessment: Readings from educational leadership*. Alexandria, VA: Association for Supervision and Curriculum Development.

Center on Learning, Assessment, and School Structure (CLASS). (1993). *Standards, not standardization*. Vol. 3: Rethinking student assessment. Geneseo, NY: Center on Learning, Assessment, and School Structure.

Darling-Hammond, L. (1988). Policy and professionalism. In: A. Lieberman (Ed.), *Building a Professional Culture in Schools*. New York: Teachers College Press.

Deming, W. E. (1986). *Out of the crisis*. Cambridge, MA: MIT Center for Advanced Engineering Study.

Dewey, J. (1938). *Logic, the theory of inquiry*. NY: H. Holt & Co.

Dossey, J., Mullis, I., & Jones, L. (1993). *Can students do mathematical problem solving?* Princeton, NJ: Educational Testing Service. (ERIC Document Reproduction Service No. ED 362 539)

Harnisch, D. L. (1994). Performance assessment in review: New directions for assessing student understanding. *International Journal of Educational Research, 21*, 341–350.

Harnisch, D. L. (1995). *A practical guide to performance assessment in mathematics*. Illinois State Board of Education, Springfield. [Printed by the Illinois State Board of Education under the title of "Performance Assessment in Mathematics: Approaches to Open-Ended Problems"].

Harnisch, D. L. (1997). Assessment linkages with instruction and professional staff development. In: P. Thurston & J. Ward (Eds), *Advances in Educational Administration*, 5 (pp. 145–164). Greenwich, CT: JAI.

Harnisch, D. L., & Mabry, L. (1993). Issues in the development and evaluation of alternative assessments. *Journal of Curriculum Studies, 25*(2), 179–187.

Harnisch, D. L., Tatsuoka, K., & Wilkens, J. (1996, April). *Perspectives on reporting SAT proficiency scaling results to students and teachers*. Paper presented at the Annual Meeting of the National Council on Measurement in Education, New York, NY.

Herman, J. (1998). *The state of performance assessments*. The School Administrator. AASA Online, December.

Izumi, L. T. (1999). *Developing and implementing academic standard: A template for legislative and policy reform*. San Francisco: Pacific Research Institute for Public Policy.

National Assessment of Educational Progress. (1993). *Setting performance standards for student achievement*. Princeton, NJ: Educational Testing Service.

National Council of Teachers of Mathematics. (1998). *Professional standards for mathematics*. Reston, VA: Author.

Shavelson, R. J., Baxter, G. P., & Gao, X. (1993). Sampling variability of performance assessments. *Journal of Educational Measurement, 30*(3), 215–232.

Shavelson, R. J., Baxter, G. P., & Pine, J. (1992). Performance assessments: Political rhetoric and measurement reality. *Educational Researcher, 21*(4), 22–27.

Steffy, B. E. (1995). *Authentic assessment and curriculum alignment: Meeting the challenge of national standards*. Rockport, MA: ProActive Publishing.

Wiggins, G. (1998). *Educative assessment: Designing assessments to inform and improve student performance*. San Francisco: Jossey-Bass.

APPENDIX

A SAMPLE OF BENCHMARKS FOR LANGUAGE ARTS AT GRADE 8 (SOURCE: HONG KONG INTERNATIONAL SCHOOL, DEVELOPED BY SUSIE HEINRICH)

Benchmark	Performance Task
• Using a variety of media resources, students will present formal speeches with an introduction, body, and conclusion	• Students will put together a multimedia presentation on conflict
• Students will present ideas and issues resulting from discussions in informal class settings	• Students will present their ideas from their Outward Bound experience in groups after discussion in a question carousel • Literature Circles completely student-led
• Students will use peers as conference partners and use their direct feedback to revise and edit a piece of writing.	• Students revise and edit through conferences in partners
• Students will be able to write their experiences in different genres including narratives, free verse poems, opinion pieces, and journal entries	• Students will write for and publish an exploration magazine based on their exploration experiences • Students will write an exploratoration piece reflecting on their name, its meaning and the experiences they've had with it.
• Students will view each other's work and their own, analyzing and evaluating their success	• Students will use exemplars and real life work to create rubrics and assess their own work.

Benchmark	Performance Task
• Students will assess their own writing strengths by using analytic rubrics.	• Students will present a writing defense. • Students will select their best pieces of writing.
• Students will use periodic self-reflections to assess strengths, weaknesses, and goal settings.	• Portfolios and selection of writing pieces.
• Students will be able to use a variety of note-taking strategies while reading and listening.	• Students will make study notes from their USH textbook • Students will take notes on note cards for research projects
• Using a variety of media, resources and technology, students will write a research paper inclusive of an introduction, body with transition statements and conclusion relating to a thesis statement. • Students will elaborate on their ideas and support their ideas and opinions with facts. • Students will compare information from different sources and texts. • Students will be able to distinguish between relevant and irrelevant information in a multitude of resources. • Students will be able to distinguish between fact and opinion in their own writing and in a given text.	• Students will write a research essay on an explorer and their contributions to the world. • Students write an essay on an explorer comparing sources and evaluating the information found.

Benchmark	Performance Task
• Students will correctly cite a variety of print and nonprint sources and present them in a correct bibliography.	• Students take part in Word Sleuth where they use reference material to discover a word.
• Students will define and recognize the elements of short stories including theme, plot, setting and character development, conflict and will critically respond to the issues raised in various stories. • Students will be able to identify literary elements/devices such as similes and metaphors, theme, dialect, foreshadowing, and point of view.	• Students will read and discuss a variety of short stories, leading to a final assignment of writing their own short story and publishing a short story anthology
• Students will generate interesting questions to be answered during or after their reading	• Literature Circles Discussion Director
• Students will understand stories and expository texts from the perspectives of the attitudes and values of the time period in which they were written.	• Literature that is integrated with USH such as Johnny Tremain and Killer Angels
• Students will respond to a variety of adolescent texts using reading logs and letters and written and spoken dialogue with peers.	• Students write formal and informal letters to a variety of audiences reflecting on their reading and connecting it to real life • Literature Circles